**Statistics Topics**
**Salil Mehta**

Independent statistics advisor

Adjunct Professor of Statistics, and Analytics
Georgetown University

Former Department Director of Policy, Research, and Analysis
Pension Benefit Guaranty Corporation

Former Director of Analytics for TARP
U.S. Department of the Treasury

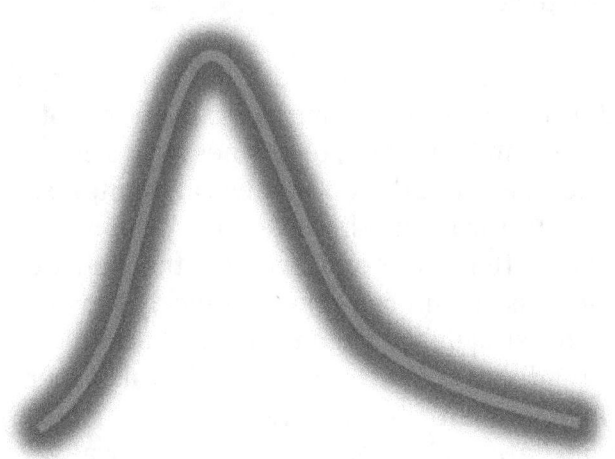

## Background

With 17 years of professional experience, I have seen the use of statistics at the edge of innovation. From top Wall Street firms, to Washington, to science and academia. My own views on how other people misinterpret statistical information are based on experience of working with many professionals globally, in a variety of work capacities. This topic is now of greater importance with the growing interest for big data and the information revolution pushing their way into relevancy, by pouring over large heaps of data, and just hoping that this will serve a positive objective. The topic is also of greater importance as nations seek to gain in their population's mathematical literacy and better compete in the current century.

My recent work experience includes running the analytics group for the Administration's $700 billion TARP bailout program. This was the highly visible capital injections into a range of industries: such as autos, insurance, banks, and housing. I am also the former department head for policy, research, and analysis at the PBGC. There I was in a highly visible position working with the agency head and managing teams of actuaries, consultants, and policy experts. All with the goal of better understanding the future trajectory of the nation's pension system.

I am currently an independent statistics advisor to high-profile corporate and educational entities. I am also a professor of statistics, biostatistics, and analytics, to business and science graduate degree students at both Georgetown, and Rutgers. This is in addition to lecturing at other leading universities prior to this year. And 18 months ago, I created the Statistics Ideas web log. This base of information has quickly shown up on 5 universities' course syllabi and central bank backlinks, and it has >9,000 readers per article. My work has also been well recognized independently by two prestigious statistics societies: American Statistical Association, and the Royal Statistical Society. And there is more to come, for example a leading Society of Actuaries publication, and other academic reviews are in the works. Additionally the Kids' statistics internet-playground is being built in 2014.

What I find amazing about the probability models that we use today is that they have had their origins, many hundreds of years ago, in similar applied settings from gambling (think Monte Carlo instead of Wall Street), to manufacturing. Please join me with this book, and explore a range of statistics topics, highlighted in a clean and fun way.

I was first motivated to write this book more than a decade ago while a graduate student in Harvard's statistics department; but only in the past couple years did everything bang together. There were many professionals, and family who have helped shape my career in the past half-decade. Every person who has hired me during this time was my earliest and most significant risk takers. Others have provided support, both big and small (though it was never that small). I received detailed counsel and hospitality from Myles Thompson of Columbia Press, and Vincent Burke of Johns Hopkins Press. In recent years, several media professionals were instrumental in bringing visibility to my work. Namely Jeff Sommer of The New York Times, Tom Keene and Betty Liu, both of Bloomberg, Lauren Foster of CFA Institute, and professors Stock and Watson. In academic and publishing spheres, Dave Bayer, Stephen Dubner, Lisa Gallagher, Stephen Mihm, have all provided great advice. The editors of arXiv, SSRN, and RePec have provided academic math and economics platforms to showcase my statistics research, and Georgetown, and Rutgers were both great in allowing the book to also be integrated with class. And I am grateful for my wife, Charu, for her constant encouragement to focus on the things that matter. Thank you as well to the many who have provided valuable counsel, and whom I have missed in this list.

It is for the people above, as well as my own desires, that I am donating half of this book's income to philanthropic efforts. So that others can enjoy the same helping hand afforded me.

**Table of Contents**

# Chapter 1: Probability theory

We see and use the word "probable" a lot. It is used to express different things. We see it used, for example, to describe that an event will likely happen. Other times it is used to give a more precise certainty over an event's occurrence. And sometimes we actually don't think an event is likely to occur at all, but we want to use the term to try quantifying this low likelihood. In this chapter, we discuss different ways one can use the term probable, and how it connects to the formal definition of probability. Here, as well as throughout the book, we will walk through examples. This way we can visually think through topics of probability and statistics ideas, while reading stories.

We will start with this sentence below, to help understand the different ways one might interpret the meaning of the term probability. What would you make of this sentence?

*The probability of finding a job after college graduation is low, because the probability of being unemployed after college graduation is high.*

At first blush, one might think -based on the wording- that the probability of finding a job is less than 50%. Or that the probability of being unemployed at graduation is higher than 50%. Without a reference, it is common for many to use 50% (half) as the cut-off between a "low" probability and "high" probability. Note that nowhere in that sentence do we mention the timeframe of the comment, nor the economic region where it applies. So even a simple factual sentence leaves room for ambiguity, even though the formal definition of probability makes us want to pin things down to one indisputable value.

Now let's look at another example. Is a 1-in-6 odds, high or low? One might again think it is low, but the result should be dependent upon the utility behind the context. Say that this is a deadly Russian roulette game, with a modified pistol pre-loaded with generally one bullet in a 6-chamber pistol (and then the pistol is sealed so that one ca not actually see the number of bullets in the chambers).  The manufacturer certainly tries, as all manufacturers' do, to make sure the process is completed to the exact specifications, but there are often some random errors (plus or minus) that are tolerated.  We see this on wide display right now, as Mary Barra defends GM's usage of supplier-made faulty ignition switch parts that not only resulted in a number of deaths, but also were known a-priori to not to meet internal equipment specifications.

Returning to our pistol example, sometimes because of minor errors the pistol manufacturer doesn't pre-load any bullets, or sometimes they pre-load two bullets.  Either way, the 1-in-6 odds certainly now appears high, since we are using the context of death to evaluate these probability odds.

Also, we never know it is a 5-in-6 chance of survival for sure.  The terms "probability", "confidence", and "likelihood" are often interchanged, even though these are distinct definitions.  The odds we have been referring to just a probability, though if we conditionalize those odds, then we can use the term likelihood.  And if we put an error estimate on those odds, then we would use the term confidence.

For probability, we can define this as the general portion of event outcomes, where a specific event occurs.  As an example, if 2 out of 10 college graduates do not find work, then this would be considered a 20% probability.  If there were a 1-in-6 chance of death from Russian roulette, then this would be a 1/6 probability.

Event probabilities can only have a value between 0 and 1, or between 0% and 100%.  Now we analyze probability theory from a theoretical point of view, while we will look at statistical models with a more empirical vantage point.[i]  A bedrock concept of probability that we use throughout this chapter is the "total probability theorem", which states that a total set of mutually exclusive, and completely exhaustive (MECE) events equals the total probability of one.

We'll see this utilizing what is known as a Venn diagram, in Figure 1.1, to illustrate the concept of MECE. The illustration shows car ownership among adults, and the Venn space is always equal to 100%. The probability of owning a domestic car is 50% (event D+B). The probability of owning a foreign car is 30% (event F+B). The probability of not owning a car is 30%.

What is the probability of owning a domestic car, a foreign car, or no car at all? It is not 50%+30%+30%. A trivial reason for this is that this sum is 110%, which is greater then the total possible of 100%. A more robust answer is that these events are not mutually exclusive, since we can jointly own a domestic and a foreign car. So we need to disjoin our double counting of event B when summing together our domestic and foreign car ownership probabilities.

Also note that event C is a complementary event, implying any remaining probability, after all of the circled events are accounted for. So the entire Venn space, or the events plus the complement, again sums to one.

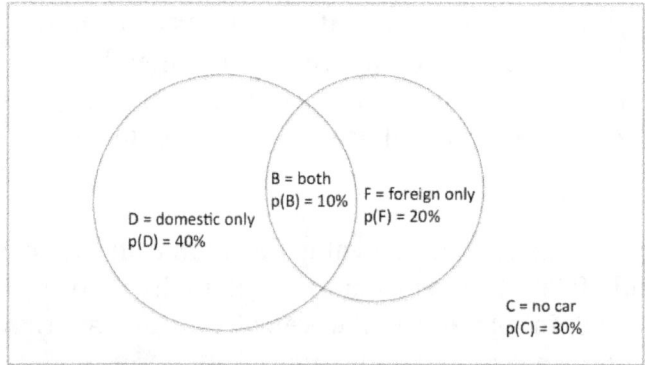

**Figure 1.1**

8

Adding together the multiple probability scenarios can be done, with varying degrees of complication, through the Venn diagram. The easiest example is if we wanted to think about the probabilities in Figure 1.1 as being associated with car selling events, at any given time. Say it shows that the probability a new customer would enter a showroom on a given day and purchase at least one vehicle is 70%. What would be the probability of this occurring two days straight with two new customers? If these were independent events, such that one event's probability doesn't influence the other, then we can simply take 70%*70%=49%. Note that the units are never "%%" (some awkwardly name this unitless.)

If the events instead were dependent, such that new customer on the second day mimics the behavior the new customer he or she saw from the day before, then the above probability question would stay 70%. In both the cases we just discussed, we notice the sub-additivity probability principle, which states that the probability of a combination is less than or equal to the probability of their weighted sums. In this case: ½*70%+½*70%.

Returning to the topic of independence, there is a concept termed "convolution", which applies if there are multiple ways to sum to a particular outcome. For example, say we want to know the probability of at least two a domestic and a foreign car being sold, when there is one customer in each of two days. Here we know that the probability is at least the sum of three possibilities. The first is a 10% probability of a customer purchasing both at the same time. Then there is a probability of one being sold on one day, and then at least no cars being sold the following: 40%*(100%-70%)+20%*(100%-70%). The (100%-70%) is for the 30% complement of no cars being sold. And then there is the probability of just two cars being sold at the same time on day two: (100%-70%)*10%. So we have 10%+(12%+6%)+3%. Or at least 31%.

This Venn diagram begins to get far more complicated when we are dealing with nuances from a large number of trials (e.g., the chance of selling three cars in more than four days). So we use a multimonial model to help tease out the mathematics quicker. The "monial" suffix refers to the number of the outcomes. The easiest and most popular version of this is the binomial model, which is a repeated process with the same two outcome possibilities (e.g., the outcome of a fair coin toss).

In order to explore the binomial model, we will first go through a simpler version of it, named the Bernoulli probability distribution. We explore it in greater detail in Chapter 3, but for now we will show an example of a "fair" coin toss. A fair coin toss has an equal probability (i.e., 50% in this case) of landing on heads, and of landing on tails.

We now expand this Bernoulli event by tossing the coin a few times, instead of only once, and recording the chance of only landing on heads given the number of tosses we perform:

| | | |
|---|---|---|
| *1 toss* | $= 50\%^1$ | *= 1/2 probability of showing heads* |
| *2 tosses* | $= 50\%^2$ | *= 1/4 probability of both showing heads* |
| *3 tosses* | $= 50\%^3$ | *= 1/8 probability of all three showing heads* |

One can imagine this list continuing to grow infinitely large. In the case of two coin tosses, what is the probability of the final tally being one heads, and one tails? We see from earlier that there is a ¼ probability of both tosses resulting in heads, so by symmetry we have the same probability for both tosses resulting in tails. Therefore the complementing probability that remains is for one heads and one tails, or $1-¼-¼=½$.

This coin tossing experiment can be expanded to many more trials, and this is what is the binomial distribution. And in doing so, the mathematics would become more difficult. Say for example, what is the probability of getting two heads and three tails, after performing five coin flips? What if the coin was not fairly balanced between heads and tails? For these advanced questions we must turn to combinatorial math to short cut a solution. We show an example below, for five coin flips.[ii]

*H*          = *coin flip shows heads*
*T*          = *coin flip shows tails*
*n*          = *number of coin flips*
*m*          = *placeholder for # of heads shown in n coin flips*
$_nC_m$          = *combination of ways to attain m heads in n coin flips, where the order doesn't matter (e.g., H,T,T,T is not a different combination from T,T,H,T)*

$$= n! \, / \, [m! \, (n\text{-}m)!]$$

*p(m)*
$$= {_5C_m} \, (\tfrac{1}{2})^5$$
$$= {_5C_m} \, (\tfrac{1}{2})^{5\text{-}m} \, (\tfrac{1}{2})^m$$
$$= {_5C_m} \, p(T)^{5\text{-}m} \, p(H)^m$$

*p(0 heads in 5 flips)*    $= 5! \, / \, [0! \, (5\text{-}0)!] \, (\tfrac{1}{2})^5$
                      $= 1/32$, *or*       •

*p(1 heads in 5 flips)*    $= 5! \, / \, [1! \, (5\text{-}1)!] \, (\tfrac{1}{2})^5$
                      $= 5/32$, *or*       •••••

*p(2 heads in 5 flips)*    $= 5! \, / \, [2! \, (5\text{-}2)!] \, (\tfrac{1}{2})^5$
                      $= 10/32$, *or*       ••••••••••

*p(3 heads in 5 flips)*    $= 5! \, / \, [3! \, (5\text{-}3)!] \, (\tfrac{1}{2})^5$
                      $= 10/32$, *or*       ••••••••••

*p(4 heads in 5 flips)*    $= 5! \, / \, [4! \, (5\text{-}4)!] \, (\tfrac{1}{2})^5$
                      $= 5/32$, *or*       •••••

*p(5 heads in 5 flips)*    $= 5! \, / \, [5! \, (5\text{-}5)!] \, (\tfrac{1}{2})^5$
                      $= 1/32$, *or*       •

Notice that the probability pattern went from a triangle (¼,½,¼) when we had only two coin flips, to now more of a bell-shaped curve ($^1/_{32}$,$^5/_{32}$,$^{10}/_{32}$,$^{10}/_{32}$,$^5/_{32}$,$^1/_{32}$) when we have five coin flips?

The history of mathematics goes back thousands of years, and pre-dates some of the earliest popular science discoveries. We can see evidence of this from ancient Chinese art, all the way across to the Great Pyramid of Giza. Probability is a relatively newer field, with popular research done many hundreds of years ago in Europe. So in this context we see that statistics' "normal" distribution goes back only a briefer amount of time, when scientists were approximating natural phenomena through this bell-shaped distribution. An early 19th century scientist named Carl Gauss was able to integrate this shape into a number of real world applications, and thus the popular distribution is synonymously known as "Gaussian" as opposed to "normal".

In order to achieve a normal distribution however, a large sample size of greater than 30 is required. This is also expressed as the "limiting" approach of the binominal distribution being the normal distribution. What we are stating with this approach is that one distribution begins to mimic the other, in this case only when one has a large sample size.

Of course this works better, the closer the binomial model's event probability (e.g., the result of each coin toss) is to 50% to begin with. Here and throughout the book, we build upon our limiting distribution tree, as we learn new distributions and how they connect to one another. So we begin our probability tree here:

*limiting distribution = l.d.*

*Bernoulli* →$_{l.d.}$ *binomial* →$_{l.d.}$ *normal*

Using limiting distribution has some benefits, for example providing a more precise answer to theoretical model. The probability of getting greater then 75% of coin tosses resulting in heads, for example, is better answered with a simulation consisting of a hundred coin flips. Though the latter is better than say a simulation with just several coin flips.

There are other probability models besides the discrete binomial, or the continuous normal distribution. Discrete means that there are a finite number of outcomes (e.g., the number of daily chicken births at a farm), while continuous has an infinite choice of outcomes (e.g., the daily amount of cow milk produced at a farm).

Another popular probability model is known as the Poisson distribution. In the 19th century, a zealous young mathematician, Siméon Poisson, revolutionized branches of physics while studying under two famous mathematicians and astronomers: Pierre-Simon Laplace, and Joseph-Louis Lagrange. In a field closer to earth, he left a narrow and important mark in the probability field. A discrete model that connects the probability of event counts, with the exponential time in-between those events (for physical processes, and well beyond).[iii]

For example, we can think of the total number of snowstorms through a given time period. Snowstorms can sometimes cluster with many of short-frequencies packed into a season (e.g., the U.S. in 2014), or other times go multiple years with no major snow storm (e.g., the U.S. in 2011 and 2012).

The Poisson model connects these ideas together, in a positive-only outcomes model. For example, the number of storms per season can not be less than zero. Because the lower-bound cases are so small (e.g., zero, one, etc.) we don't need as many of these large interval events covering any time period.

To illustrate the Poisson model idea, take a look at a hypothetical hurricane event distribution in light grey, of Figure 1.2 below.

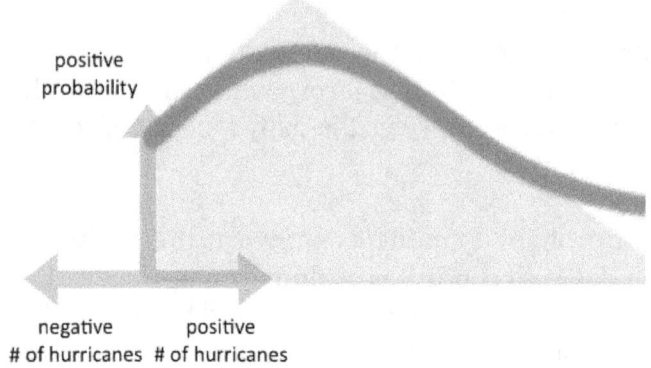

**Figure 1.2**

While the Poisson is not a perfect fit, it has some distinguishing qualities that are ideal for this frequency chart. We'll explore those qualities now.

One thing we might notice is that the Poisson has a "thicker" right tail than the distribution we seek, and in theory it extends forever. Other characteristics that are worthy to note now is that the Poisson can only take on non-negative values, which is ideal for reporting hurricane event counts. Also a neat property of the Poisson distribution is that once we have an expected average for our variable, there are no other parameters to be solved to understand the full distribution.

We'll discuss the concept of parameters in greater detail, later in the book. For now though let's note that parameters are key attributes that can solve for what a random variable distribution outcome looks like (e.g., the probability of seeing no hurricanes in a period, or one hurricane, or two, etc.) These terms will be clearer as we repeatedly reinforce them through the book.

If we had used a binomial model instead, then we would have need two parameters to distinguish among them: (A) the probability of success, and (B) the number of trials. Notice that for the clumpy shape of the Poisson distribution, the probability of being within one less than the average event (or at least zero, if the average event is less than one) is generally higher than the probability of being within one greater than the average event. So if the average number of hurricanes per year is two, then we might expect a higher probability of one hurricane versus the probability of either two or three hurricanes.

We also note that the Poisson process mathematically connects the frequency of events, and the time in-between events. For example, a low frequency event (e.g., 1 hurricane per year) doesn't occur as often, and also has a rare large amount of time in-between events (e.g., four quarters). On the other hand, there are many events that exponentially occur in a short amount of time in-between events (e.g., one quarter), but they are reduced in the count for a high frequency probability (e.g., 4 hurricanes per year). This identical new distribution from one event to the next, in the exponential distribution, is an exotic form of what probabilists term "independent and identically distributed" random events. Or "i.i.d." for shorthand. We show in the earlier endnote that economic models do not use practical sense on these extreme frequency assumptions, and hence give a slightly higher bias on the length of economic cycles.

Returning to our hurricane example, here is a hypothetical 5-year period of annual global storm count data (with an average of 4 hurricanes per year):

| Year | Annual storms | Deviation from avg. | Deviations$^2$ |
|------|---------------|---------------------|----------------|
| 1 | 4, or •••• | 0, or | 0, or |
| 2 | 4, or •••• | 0, or | 0, or |
| 3 | 6, or ••••••• | 2, or •• | 4, or •••• |
| 4 | 2, or •• | (2), or (••) | 4, or •••• |
| 5 | 4, or •••• | 0, or | 0, or |

avg. →4, or  ••••

avg. deviations$^2$ =                    variance →   ~2, or ••

We show the squared variations about the average, for all years.  This is the variance.  For the Poisson distribution, the variance is also expected to equal to the average.  Sheldon Ross' probability book has a nice proof of this.[iv]  Note that the "~" symbol means "about".

We'll note that in our earlier example that the average of 4 is not ~2, however.  A larger sample size would help allow these statistics to align better, however an important concept we need to know when using statistics is that we usually do not see what we might precisely expect for a trial.  Though it is still correct if we should see this expected value over a lengthier time, on average.  We'll discuss the size of representative samples later in the book.  Though it's fair to say we need to be cautious in real world application where we often only have small samples (e.g., <30) initially.  We always argue in this book that, when possible, it is better to wait for much more additional data before making a conclusion.

It is worth repeating that for statistics, one issue with anything but an extraordinarily large sample size is that the results are exceptionally fickle.  And people quickly anchor their beliefs and bias their thought patterns in preliminary results.  They are also much most confident in the results then is warranted, when instead the results can change quite much in the next sample. Recently we have seen evidence of this with flight MH370, where local governments would daily come out with new leads and an exciting frenzy would ensue until it was wrong and the process would restart the next day.[v] In reality, all the comments should have been better couched with the unreliability that the "evidence" was (e.g., two debris that can happen to float near one another at one point, somewhere in the vast ocean).

And none of what we discussed so far, mentions the higher proportional displacement possible with any data errors. For example, what if Year 1 data on the prior page should have had a value of 9, instead of being miscoded right now as 4. Well that's a problem. This would raise the overall average from 4, to 5. But this one error would also dramatically spike the variance (the squared differences from the average), from ~2, to ~6. In practice, one could "stretch" the original Poisson data in the table above, so that the new variance is a better match.[vi] And in practice, one should be super vigilant in searching out data quality issues before crunching results for a deadline.

We should note that in the insurance math of modeling total storm damage per unit of time (e.g., season, decade, etc.), we have to consider not only the quantity of storm counts, but also the damage severity cost of storms. We model severity as a second and independent random variable. The average storm damage is easy; just (average count)*(average severity).

But what about the variance about this average storm damage value? This is a trickier probability formula as the result is a non-linear function, which we later show is more complicated.[vii] We show in Formula A, in the back of the book, how our traditional Poisson properties could be used to solve for the variance of the total cost ($\sigma_{sum}^2$). We just assume the severity unwavers at a value of 1.

It is worth noting again that when looking at deviations$^2$, we often need to consider removing the power of the units associated with the numerical value. For example, if we consider a statistic that we typically see a 25% deviation from average in hurricanes in any season, what would deviations$^2$ look like? Would it be 6.3 percent$^2$ or just 6.3%? After all, the original expression of hurricanes$^2$ doesn't have a literal meaning. Nonetheless, we do need to appreciate that these unit powers exist, particularly when the individual would need to exert additional calculations beyond the one that got us to this point. For example, we often take the square root of the variance in order to get the standard deviation, and need to get to an answer with the units of hurricanes and not $\sqrt{hurricanes}$.

We do not have the ability to go into great detail on all of the probability distributions available in the field. Most of them are anyway too exotic to ever normally come across for a lay individual in their lifetime. Some distributions also lay dormant for decades, and then suddenly become popular in unintended ways. After the 2008 financial crisis, more risk models began using the science of extreme value theories that have been somewhat unnoticed in finance for decades, and which utilize the well-developed generalized Pareto distribution.[viii]

Some other probability distributions that would be highly unlikely for non-scientists to come across, include the Beta, and Weibull models. These distributions have thinner-tails, for example the Beta that is essentially a fancier version of the uniform-spread distribution. And the Weibull distribution is associated with yet another family of distributions that are popular only for quantitative professionals. Those distributions include the negative binomial, and geometric distributions (these are popular in science and certainly taught to those students). We'll see in Chapter 9 that these distributions just mentioned in this paragraph are some of the most popular distributions on the internet.

What's interesting about the some of the most popular theoretical probability models that we take for granted and often use today, is that they were developed centuries ago in Europe in order to respond to not just to gambling problems, but to mostly other pedestrian phenomena (e.g., everything from the characteristics of children's heights, to a beer plant's output).

Yet other theoretical distributions can be created, and there is no reason to consider the ones that currently exist to be the ultimate list. The key of course is not whether it is useful (as noted sometimes this can't be proven until well after it is discovered), but rather the simple test of whether a distribution is actually completely unique, and not just a dependent mixture of existing models.

Here's an analogy. Let's consider a "probability model" to be a unique beverage from a restaurant fountain that currently only dispenses two options: one for lemonade; the other for iced tea. Lemonade and iced tea would both be considered unique models, as one can't do something to one, in order to create the other. But simply mixing different portions of these two models together would not be unique. The Arnold Palmer beverage is therefore not a unique model, since it is simply a mixture of two existing models. Chocolate milk on the other hand could be considered a unique model, since it can not be created through the mixture of the current exiting models. We can translate from this analogy that it is much more difficult to create new drinks as revolutionary as chocolate milk, as it is to create mixed beverages, such as the Arnold palmer.

Now let's now discuss the difference between mixing two probability distributions, and taking the average of them. To start, mixing two random variables retains the original range of both variables treated as one (with a maximum of both being the maximum of the mixture, and the minimum of both being the minimum of the mixture). This is mathematically easier since we are simply recasting the original variable values. On the contrary, averaging two random variables can morph the range. This is because of the more complicated concept of convolution we described earlier, and we go through a simple example now.

Say that we toss two different coins (since we only have two variables, convolution simplifies to just averaging the output from the paired coin toss). The first coin has 0.0 and 1.0 on its faces; the other coin has 2.0 and 3.0 on its faces.

*Randomly pick one of the coins and flip it. The mixture of these two coins then results as follows:*

| Event outcome | Mixing result | Probability |
|---|---|---|
| 0 | 0.0 | 25% |
| 1 | 1.0 | 25% |
| 2 | 2.0 | 25% |
| 3 | 3.0 | 25% |

*Next, flip both coins and average the face values. The average is as follows:*

| Event outcome | Averaging result | Probability |
|---|---|---|
| {0,2} | 1.0 | 25% |
| {1,2} or {0,3} | 1.5 | 50% |
| {1,3} | 2.0 | 25% |

The enumeration above works easiest when dealing with two Bernoulli distributions (e.g., two coin tosses), or two Poisson distributions (e.g., the passing of either a freight train or a passenger train both sharing the same bridge track). But this enumeration quickly gets more complicated once we deal with many additional trials, or different types of models. Here the convolution principles would apply (e.g., say we are looking at the sum of three dice rolls).

Despite having probability tools to look at random phenomena, there will always be some unknowns about how to describe an observed event. In the first half of 2013 the U.S. financial markets enjoyed record streaks of up-days. Were these random events independent of one another? And what is the confidence surrounding the future probability on any estimate?  Simply appreciating that these unknowns are out there, which make difficult bridging the gap between probability and observed data, can assist in thinking about the value-add and the limitations of statistics.  We'll have to keep these difficulties in mind when considering the statistics topics developed throughout this book.

## Chapter 2: Parameters

Parameters are an efficient way to describe a random distribution. Think of describing what a Boeing's Dreamliner basically is, to someone who has never even heard of it. Except the challenge is to describe it with the most efficient (least necessary) number of words.

One way to describe it is to discuss the Dreamliner's specifications, such as the length, width, and weight. And a more nuanced approach is to instead discuss the engine size, number of seats, and type of onboard communications equipment. The point here is that parameters provide a small number of independent characteristics, which narrow the range of possibilities of what the plane can be.

The key here is to also not describe what is not necessary, and this differs from one probability model to the next. In the Dreamliner example, once we provide the full length of the plane, we don't need to establish another parameter for the half-length of the plane or even the length of any seat on the plane. The latter two dimensions simply being not as necessary as the former, in order to understand the overall plane specifications. Of course describing the specifications of something simpler, such as a single playing card, would require less parameters than describing a Dreamliner.

The most common parameter for a distribution is the average. The second most common parameter is a measure of scale, for example the dispersion of outcomes relative to the average. Among many examples of how people shy away from looking at dispersion is a recent popular Harvard study, which concluded that the Social Security Trust Funds would turn insolvent earlier than expected. The media covered the provocative headline of that the funds should turn insolvent sooner than expected. But no one went on to discuss the equally relevant question how the Harvard study's total modeling uncertainty was already well known by traditional insolvency analysis at that time. Of course this imbalanced reporting of the study's sensitivity to other risk factors proved important, as a later new, official government estimates provided results in the opposite direction of that Harvard study.[ix]

One can and should look at the idea of dispersion a number of ways; the variance statistic is not the only way. For example, look at values for specific ranks of the sorted distribution. To see how this is done, note that in the summer of 2012, the Congressional Budget Office released a report showing the household income distribution in the U.S., during a 30-year period through 2009.[x] They stated that the top-1% of income households, had a cumulative growth in after-tax income of nearly 300%. Meanwhile the bottom-20% of income households had a slower growth statistic of about 40%. Note that instead of the two parameters of average and standard deviation, here we instead looked at data values at two disperse ends of the sorted distribution and were left with the equal challenge to imagine the rest of a smoothed distribution accordingly. This idea of looking at different parameters for the same topic is something that has become more popular since 2008, by global insurance governing bodies advocating ideas on modern risk management.[xi]

There are other important considerations with parameters, including how they might be impacted from any outliers. Outliers, which we will explore more deeply later, can for now be thought of as data values furthest from the rest of the distribution. Say one has two sets of data:
{-1,0,1} and {-1,0,100}

Then both sets have the same middle value of 0. But the presence of an outlier in the second set (i.e., 100), forces the average value to jump from 0 (also the median in both sets) in the 1st set, to 33 in the 2nd set.

In addition to location and the scale, another question we would want our parameter(s) to be able to answer is the shape of the distribution. Say that our models were meant to articulate two countries' shapes. And say that we only had two parameters measured so far, the border length and area, both of which are 10,000 miles$^2$ and 5,000 miles$^2$, respectively. Would you therefore conclude the shapes of the countries are identical? Of course not. These parameter values could have been calculated from a country that is the shape of a triangle, barbell, quadrilateral, etc. So in the case of some distribution models, we still need more parameters to capture a decent amount of information about what the model is.

Another consideration for selecting parameters is the degree of uncertainty in the distribution. If for example, we take a model of rebels and protesters recently in the Middle East to be democratic-friendly, but instead we have mis-sampled this group, then it is very likely that we will not be able to represent a clear picture of the underlying population (population here refers to the complete probability distribution, as opposed to a sample). We have seen this in a number of countries there, where during civil wars it was impossibly confusing for outside governments to know what the alliances and motivations were of people, simply by sampling those in one-time rebel cities. We'll reference the recent example of Damascus, in Chapter 5.

A final thing we should consider with parameters is that sometimes we may want to focus our analysis on only a subset of the distribution. In extreme value theory used for modern risk modeling, we may only want to look at the worst-1% of market movements. Or for the hurricane distribution model, we may use something similar to a normal distribution, except first truncating values below zero. For admissions staff in local gifted and talented schools across the globe, they need a way of assessing the top 10% of applicants in order to determine who should advance to final round interviews, using methods different than if they were looking to curve rank all 100% of the applicants.

In summary, we look at parameters to be able to efficiently describe the shape, scale, location, and uncertainty of the distribution. Sometimes this can be done in just a couple parameters; sometimes more parameters are needed.

Now we go through some examples to show how mathematically difficult it is to isolate the parameter values that should be fitted against a sample of data. The first example concerns a metal detector company that advertises a product, which can typically find three coins every bi-hour (i.e., every two hours). After purchasing this product, we discover with it one coin in the first two hours, and then find three additional coins in the third hour. So in total, four coins were found in the first three hours. Did the coin detector perform as advertised?

This example fits the Poisson probability model; so only one parameter is needed to solve for both the average and variance. Here the advertisement says the frequency is 3 coins every two hours. But as is typical with many real-world examples, the parameter value isn't known. Or it must be tested from a sample, as we are about to do next.

One approach to estimating the parameter from the sample is to rationalize. This approach may work well for a Poisson model, if the amount of time is very high (e.g., much more than eight hours in this case), and the frequency of events occurring per unit of time is low. This is not the case here, but we show the rationalized math anyway as [4 coins/3 hours]*2 hours. Or 2.33 coins found per bi-hour.

We noted that rationalizing works best for a Poisson distribution when the sample time overwhelms a small sample average. In this case, the Poisson distribution also approaches the characteristics of the normal distribution via the binomial approximation. The connection between time and Bernoulli trials is also shown more precisely, in our later conversation on stochastics. And this connection also helps derive phenomena we see, such as the omnipresent Golden ratio $\phi$ (1.618...) observed in nature. Here now is our expanded distribution mappings. We will **bold italics** only the new add-ons to the mapping:

l.d.      *= limiting distribution*

*Bernoulli* $\rightarrow_{l.d.}$ *binomial* $\rightarrow_{l.d.}$ *normal*
**normal** $\leftarrow$ *l.d. with expected fit*$_{\to\infty}$ ***Poisson*** $\leftarrow$ *l.d. with sum of i.i.d.* ***Poisson***

But what do we do in the many cases where we do not have enough information to simply rationalize the parameter fit? In these more complicated, yet more typical cases, one technique is known as the method of moments (MOM), while another is the maximum likelihood estimator (MLE). We'll discuss now the similarities and differences of these two approaches.

The MOM align as many moments of our sample, as there are parameters needing to be solved. Moments, which we'll describe in greater detail in Chapter 3, refer to the number of orders to describe the sample distribution. We can think of the average as the simplest order calculation, while the dispersions[2] (e.g., variance) could be thought of as the second order. We can also think of the dispersions[3] to be a subtle measurement of overall model's shape tilting or leaning, to either the right or left; thus this could be thought of as the third order.

The Poisson distribution is defined by only one parameter, so only the basic, first moment is needed: the sample average. We avert the proof that for the Poisson, the average of $X$ is the parameter fit.[xii] But the key calculation here is that 4 coins in 3 hours would again equal 2.33 coins per bi-hour. Note a random variable usually is denoted with a capital letter, in **bold italics**.

The MLE approach differs in that it seeks to tease out the most likely yet-to-be-solved parameter $\lambda$, such that the joint probability (i.e., likelihood) across the entire range of sample results is highest. The set-up can be seen in Formula B.

For the Poisson example before us, the MLE approach results in same 2.33 coins per bi-hour result. For the Poisson, rationalizing, the MOM, and the MLE all provide the same $\lambda$ value for the parameter.

Still, the MLE is a different method to approximate the model parameter, and while it was not true in this case, sometimes both the MOM and MLE can be quite sensitive and also both can be imprecise. In some cases, one should attempt to still look at both estimate techniques, and consider both the results. Though for some cases only, such as the prior Poisson example, we saw that the results will be the same regardless.

Here are other specific distributions where the results would be the same: normal, exponential, gamma, and negative binomial. Also in some other cases, only one estimating technique is plausible to solve anyway, in which case the debate over MOM or MLE is moot.

Now we'll go through an example, where we solve for MOM≠MLE. Say that an archer shoots three arrows towards a wall, which has a large circle painted about the target center. The archer only aims to strike anywhere within the circle, except for the circle's center itself.

After a night of rain, the next morning we see the large painted circle is completely washed away. All that remains is the target center, and the arrows that impaled the wall around it. These arrows are at the following distances from the target center: 2", 4", and 4". Question: how large was the washed-away circle?

First let's look at the MOM approach, and we note that we can use a uniform distribution model from distance 0, to an unsolved distance parameter $\theta$. The average of this uniform model is $\theta/2$, and so this would be needed to match the sample average distance of $(2+4+4)/3=3.33$", or $\theta=6.67$".

Without delving into the proof, the MLE technique for a uniform distribution, with a zero lower bound, is simply the maximum sample distance. This approach gives little weight to the idea that the archer aims within the circle, but rather that his or her most deviant shot will impale the wall near that large circle itself. So our $\theta$ in this case would be 4", which is less than the 6.67" using MOM. Which value seems more appealing to you? The question in statistics is sometimes how to make a qualitative assessment from different, yet equally correct, solutions.

Of course the quality or confidence in the results from our MLE approach would be enhanced with a greater sample size. This allows for a higher chance for the parameter $\theta$ to better converge on the circle itself. The quick take from Formula B2 is that the variance surrounding the MLE estimate of $\theta$ is inversely related to the (sample size)$^2$. And the standard error of $\theta$ is the $\sqrt{}$variance, or $\sqrt{0.6}=0.8$".

Let's switch gears slightly from the concept of a stationary distribution needing a parameter, to instead look at ideas surrounding conditional distributions. The remaining concepts in this chapter explore how the probability model itself dynamically changes based upon some factor or a life choice we are making.

We start with the Monty Hall problem, which is a probability puzzle based on the American television game show *Let's Make a Deal*. It is the same problem featured in the movie, *Bringing Down the House*, where an M.I.T. professor uses the problem as part of a classroom case study. It was originally posed in a letter several decades ago:

*Suppose you're on a game show, and you're given the choice of three doors: Behind one door is a car; behind the others, goats. You pick a door, say No. 1 [but the door is not opened], and the host, who knows what's behind the doors, opens another door, say No. 3, which has a goat. He then says to you, "Do you want to pick door No. 2?" Is it to your advantage to switch your choice?*

The correct solution is always to switch doors. A large number of people, including many with science PhDs, still refuse to believe this to be the correct solution (which is why they are not statisticians!) Most people instead believe either that there is a 50-50 chance now that the car is behind either of the remaining doors (Door 1, or Door 2). Or some people incorrectly believe that their case for the original Door 1 choice is now strengthened. But this game could be framed as a Bayesian problem, where we begin with the understanding that the initial probability of the car hidden behind any of the three doors was an equal 1/3. And so for the initial Door 1 choice, this probability must remain at 1/3 through the problem.

The English mathematician Bayes developed this probability law bearing his name. We will discuss it, alongside another similar likelihood idea, later in the book. The basic Bayesian idea is to update one's understanding of a distribution, based upon new information, which in this case was that the game host revealing that there was a goat behind Door 3.

Formula C takes a little bit of staring at, in order for it to make better sense, but we can think of event B as the probability of the goat originally being behind any of the doors. And event A

29

as being the "revised" probability of the goat being behind any of the doors, <u>after</u> the host reveals what's behind one door. And ultimately we know event B occurs, in that a goat is behind a door even though we don't know which.

Now let's return to the initial 1/3 probability of the car being behind any of the three doors, if we ignore Door 1 (i.e., the initial contestant-selected door), then the probability of the car being behind either Door 2 or Door 3 is 2/3. And these events are of course MECE. So now there is a 2/3 probability of the car being behind only Door 2, since the host later shows that the car is not behind Door 3.

It is not always easy to appreciate the conditional probability associated with an event, and in particular, what to condition the original probability on. For example, when looking at World Cup scoring results, we can model a myriad of complicated factors that a sports gambler would use to refine the odds of a match-up. These conditional factors could include things such as seasons, differing players, country match-ups, and weather.

In a similar complicated case where there are often not a rich set of quantitative data for conditional modeling, we can look at the impact of central bank activities in the recent years of 2008, through 2012. We can often argue our probability characteristics, confidently stating things such as the impact for any additional round of stimulus would be "different this time". And of course there is no way to know for sure, ex ante, who is correct. But the longer historical data shows that generally things are "not much different this time", on average. Even Mark Twain noted in the 19th century, "it is not worth while to try to keep history from repeating itself, for man's character will always make the preventing of the repetitions impossible."

This broader understanding not only helps us develop better model parameters, but also allows us to feel better about creating a confidence interval (CI) about our estimates. We show in a later chapter how to do this with central bank rate hiking decisions that continue to cause intrigue in the financial markets. In our next example we look at a problem similar to the Monty Hall problem, except here we need to make a decision before all subsequent layers of data are known.

Suppose at 1pm, one arrives to the cinema and there are only three show times remaining for the rest of afternoon: 1pm, 2pm, and 3pm. And we can only watch one, so we are hoping that it will be the highest-rated movie showing that afternoon.

Of a possible 5-star rating system, the 1pm show is rated 3-stars. The theater hasn't yet revealed the movie line-up for the 2pm or 3pm shows, but has stated that they will later decide simply by a coin flip. They also provide that the 2pm show would either be a 1-star or a 5-stars, and the 3pm show would be either a 1-star or a 3-stars.

So there is a cost, and possible benefit, to passing up the 1pm movie, in the hopes of watching a higher-rated movie later in the day. Many life choices are sometimes fraught over uncertainty with the future, but simply involve unemotionally analyzing and making human decisions now, without needing to have all future paths detailed out.

Let's visualize the total range of cinema options available, in Figure 2.1 below.

| Time | Movie description | Alternate show option (when available) |
|:---:|:---:|:---:|
| 1:00pm | 3 stars | |
| 2:00pm | 5 stars | 1 star |
| 3:00pm | 3 stars | 1 star |

**Figure 2.1**

31

In order to think through the answer, we start with the ultimate conditional average rating associated with the 3pm movie, and then work backwards to the 2pm movie probabilities. Finally we could assess the choice of whether to pass up or watch the 1pm movie. One can follow the equations in Formula C2, which uses similar logic to the pricing of options (e.g., a callable or putable bond).

We see from the example's solution that it is best to hold out at the 1pm showing, and then depending on what the 2pm movie is, make a decision to watch that movie or hold out again. Unlike in the type of solutions we had earlier, here at the decision start of 1pm, we still don't know precisely which movie time will ultimately be settled on (nor what movie rating will be settled). Though we will show below what our ongoing Bernoulli decisions will be based on what new information comes in over time.

See Figure 2.2 below for how the final Bayesian outcome landscape looks. We start on the left (at 1pm), and see there is no reason to watch this 3-stars movie as something better is likely ahead. Then depending on what's offered at 2pm, there is a 50% probability of watching a 5-stars movie, else we hold out and watch either a 1-star or a 3-stars movie (right side of illustration). The average is typically higher than the 3-stars offered at 1pm, though no guarantee to be higher than it.

| Stars | 1:00pm | 2:00pm | 3:00pm |
|---|---|---|---|
| 5 stars | | 50% | |
| 3 stars | 0% | | 25% |
| 1 star | | 0% | 25% |

**Figure 2.2**

In our final example we look at the topic of analyzing data when some of it is outright missing, and when it is unclear if it could ever be recovered. Perhaps the original data was destroyed; or perhaps survey participants simply refused to answer some questions. This does not deal with yet another important topic of purposefully manipulated data to bias the results. Sometimes performing statistics in your life is similar to Forrest Gump's comment on the latter: "...like a box of chocolates. You never know what you're gonna get."

Look at the New York City borough data, in Figure 2.3 below. How would one deal with the missing data below?

| Borough | Weekly wages per capita | Population indicator (>1 million) | Land area (miles²) |
|---|---|---|---|
| Manhattan | $1600 | 1 | 20 |
| Brooklyn | $800 | 1 | 70 |
| Queens | $900 | 1 | 110 |
| Staten Island | $800 | | 60 |
| Bronx | $900 | 0 | 40 |

**Figure 2.3**

While tempting, we don't want to either remove the entire Staten Island data row, or similarly remove the entire "Population" data column. These shortcuts seem safe, but do more damage then they do good. At best, we would be left with a data set that is not representative of the underlying population. And at worst, we have removed significant chunks of valuable variable data.

We also don't want to ignore the missing data, as if it didn't really exist. This would distort the underlying relationships among the otherwise paired-data, and provide a biased result that can lack clear analytical value.

The best approach is to estimate the value of the missing information. This is done by regressing on the values from the information "surrounding" it. A multivariate logistic will provide the most robust information. In other words we are looking at non-population variables, and non-Staten Island data. Versus no analysis at all, we can look at wages and population density in Manhattan, for example, provide some information on wages in Staten Island given its population density.

Formula D goes through this approach, where the last regression expression is $e^X/[1+e^X]$. This is helpful since we want our population indicator (**p** in the formulas) variable to also be confined between 0 and 1, as it is in Figure 2.3. This is then a perfect match, for this example, to the dependent probability variable that must also stay between 0 and 1 (for 0% and 100%, respectively). We continue the calculations in Formula E.

Our model tells us that a borough's estimated population variable starts at 0, and isn't cleanly dominated by either of the two explanatory variables. We see their dispersion and pattern in Figure 2.3. Even with real data here, we show the results are quite variable.

We now create a "super variable" **X**, which starts at -181. This is also the intercept, which dwarfs the wages data, even with the application of its 0.1 coefficient (the value multiplied by the variable value).

For Manhattan, as seen so far using the variables discussed, we get $e^{(-181+0.1*1600)} \approx 0$. But the land area variable, which happens to have lower statistical strength, would ultimately sway the population estimate from the "$e^{(-181 + ...)}$" expression above. Our result is **p** of about 0.1.

A design drawback to this answer is that we are matching results for two explanatory variables, from only four boroughs of data. And only one of those boroughs (Bronx) has a 0 population indicator value, so there is not enough data per these relatively large number of clumpy explanatory variables.

A second issue here is colinearity among variables, where multiple explanatory variables are actually related to one another, and therefore render the ultimate coefficient outputs to be extreme and distorted. Given these downsides, we can take in on Formula F, the output of this method to understand the missing data for Staten Island. We'll notice that the **p** estimate for the four known boroughs line-up perfectly: {Manhattan=1,Brooklyn=1,Queens=1,Bronx=0}.

But the confidence of the out-of-sample model output, such as Staten Island, will be lower. This is despite logistic regressions that minimize residual errors, directionally similar to our later discussion of simple linear regressions that minimize residuals[2].

Now we later recover the actual value for the missing Staten Island data, and see that it is really 0. So our estimate had a 0.1 error. While this may seem small, it is from the lower end of a 0 to 1 range. We should treat the **p** estimate with decent skepticism due to its sensitivity being based on a small sample of four complete borough data.

With this example, the major idea is that we leverage Bayesian statistics to better inform ourselves about a distribution, even with missing data. Instead of averting the missing data, had we dumbly imputed the "average" population value to estimate the missing value, then we'd instead have a far worse **p** estimate of ¾ for the average of: {Manhattan=1,Queens=1,Brooklyn=1,Bronx=0}.

## Chapter 3: Moments, and correlations

In this chapter, we explore how we can use higher order moments to help explain important details of a random distribution. And whenever possible, we also show how to visualize these statistics information.

In order to explore a probability distribution, we'll start with an easy example of the sum of a die roll. See the right of Figure 3.1 below. Any particular outcome of a die roll is random, but its complete outcome distribution is fixed and known a-priori, as we show below. The cumulative distribution function (referred to as cdf, or ogive) shows the probability of achieving a particular sum, or less, on a die roll. For example, one is less likely to roll die with a combined face values of five or less, then they are to roll a die with a combined face value of seven or less.

The median is middle outcome value (50th percentile cdf), which here we show is a die roll summing to 7.

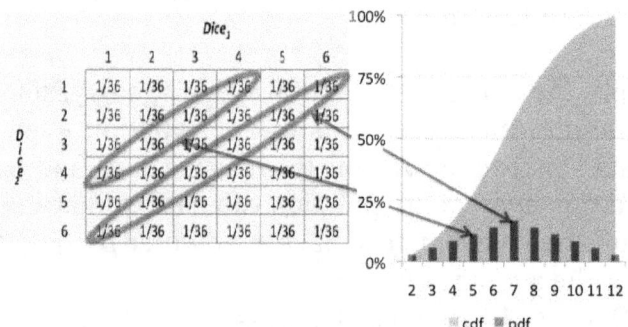

**Figure 3.1**

The probability distribution function (pdf) is the triangular shape on the right side of the illustration above. This is important since it is not a bell-shaped, normal distribution. Therefore we can as easily use the normal assumption to determine the "standard deviation" ($\sigma$). For the normal distribution, for example, we know that about 70% of its distribution is within 1 standard deviation, but there is no particular portion within 1 $\sigma$ of a triangular distribution. This is the same reason it is inappropriate to use outlier-detection rules of thumbs based only on $\sigma$ alone (e.g., some poprular tools define and propagate the incorrect notion that outliers are anything beyond 2.5 standard deviations).

We will walk through another data example so that we can further examine the descriptive properties of a random distribution. This was a component of a complex risk presentation shown to some board members of one of the largest public pensions. We have modeled a theoretical random walk, or a Gaussian model centered about a trend path. We used a 16% annualized standard deviation in simulating monthly market values, though generally we would estimate it for financial data since it is not a constant.

Look at a simulated output for 12 months, in Figure 3.2, which is from a small sample size. Note that since the sample size is less than 30, we are only looking at the higher order moments at face value. We are unable to reasonably assume the sample distribution is from, say, a normal distribution (shown in dots on the right of the chart below).

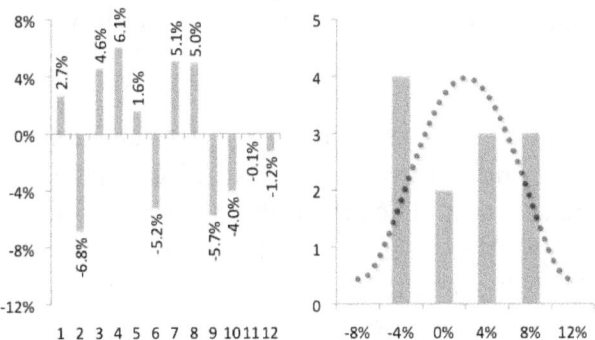

**Figure 3.2**

In order to compute the quartiles for this distribution, we assume that the 12 months sample represents a slightly broader population of 1 plus those 12 data, or implicitly ranging 13 data. Then the 1st quartile is the value ranked 25%*13 from the bottom, or -4.9%. The 3rd quartile is 25%*13 rank from the top, or 4.9%. This math only works for all but the outer quartiles (minimum or maximum data).

Now the 12-month average was nearly 0%, as this was simulated to be a "driftless" random walk. So we will take a different look at the deviations by using the absolute level of these 12 monthly changes above (e.g., 5% and -5% would both be treated as an absolute change of 5%), and then sort these absolute changes. Look at the template below to process the raw data.

| 1 | 2 | 3 | 4 | 5 | 6 | 7 | 8 | 9 | 10 | 11 | 12 |
|---|---|---|---|---|---|---|---|---|---|---|---|
| 2.7 | -6.8 | 4.6 | 6.1 | 1.6 | -5.2 | 5.1 | 5.0 | -5.7 | -4.0 | -0.1 | -1.2 |
| 2.7 | 6.8 | 4.6 | 6.1 | 1.6 | 5.2 | 5.1 | 5.0 | 5.7 | 4.0 | 0.1 | 1.2 |
| 0.1 | 1.2 | 1.6 | 2.7 | 4.0 | 4.6 | 5.0 | 5.1 | 5.2 | 5.7 | 6.1 | 6.8 |

*Rows 1-4 categories: month, raw, de-sign, ranked*

Here we see the typical monthly deviation, regardless of direction from 0%, is 3.8%. And the median of change is 4.8%, near the absolute quartile deviations from above. What's important here is that the "standard" deviation is not equal to the quartile deviations, the average absolute deviation, nor the median absolute deviation.

So what is $\sigma$? In this case we are taking the typical square of these values above, and not the "non-squared" values we used in the previous paragraph. The effect of the square approach is that we are more greatly weighting the larger absolute deviations. And this biases the square root of this underlying data towards that which are most likely outlier(s). The reason we even do this is as the linear algebra math was developed more than half a century ago, calculators, let alone computers, were not available to process absolute functions so easier manual calculations of squares and square roots gave a proxy result.  More on this topic in Chapter 8.

How can we quickly test this idea of how squaring the values would bias the result higher versus the average deviation? Look at these values {-1,-1,2}.  The standard deviation is $\sqrt{(1^2/3+1^2/3+2^2/3)}$~1.4.  This is higher than the average deviation of $(1/3+1/3+2/3)$~1.3.  For small samples of course without a large outlier type of deviation (e.g., near ∞), the average deviation could also exceed the standard deviation.

The same rules apply for median changes (the middle value of changes regardless of direction), if there are many absolute changes that are close to it and symmetrical.  Else there are no such rules, as the median doesn't reflect how the lower end of the tail distribution is.  We can have many, minute absolute changes, while at the same time we could have many absolute changes that are just higher than the median.  In our example, the standard deviation is 4.5%, between the 3.8% average and the 4.8% median deviations.

Another measure that can sometimes be used for central tendency is the mode.  This is defined as the outcome with the greatest chance of occurrence.  In the discrete example of the sum of rolling die, the mode was clearly seven. In the continuous case of the 12 monthly returns, we see it is the left-most category range from -8%, to -4%.  There are some cases, which we'll explore at the end of the book, where a distribution contains multiple probability mounds.  There we call the distribution "multimodal".  Such as bimodal, or trimodal.

When it comes to modes with continuous variables, compartmentalized into data partitions, we should be sensitive to the bins selected. We will briefly look at an example of Amtrak rail data in Figure 3.3, before returning to our normal-distribution asset return data.

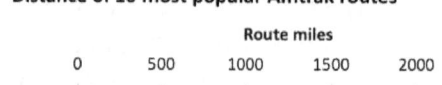

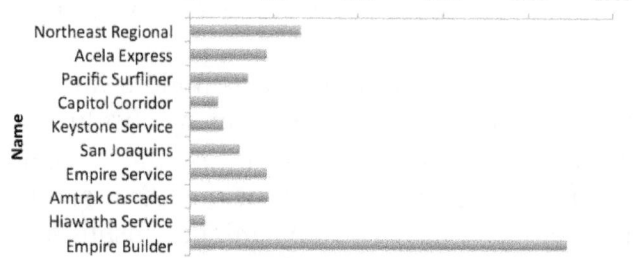

**Figure 3.3**

It is not so cut and dry to know what is the mode, and is subject to manipulation as we will now show. We could use every 250 miles for example, so that the first bin is 0-250 miles, the second is 250-500 miles, the third is 500-750 miles, etc. This would be the resulting unimodal distribution for the partitions: {3,5,1,1}. If dissatisfied, then someone could tweak the bin size to 200 miles, and silently get this bimodal distribution instead: {3,2,3,1,1}. There are many ways to partition the data so that we get a different optical sense of the center of the distribution, even though the underlying data distribution was always the same one shown in Figure 3.3. One is best served in such a quantitative descriptive data set to initially sort the quantitative data (miles), and then take another look.

40

Given the following random variable types, either in theory or a large sample distribution from the variable, we have some information on where the mode generally falls relative to the median.

| | |
|---|---|
| *median of equilateral triangular* | *= mode of equilateral triangular* |
| *median of binomial* | *= mode of binomial* |
| *median of normal* | *= mode of normal* |
| *median of Poisson* | *> mode of Poisson* |

Now returning to our asset return simulation from Figure 3.2, we note that the original population distribution was from a normal distribution but that the sample was too small to see this. In general, when more representative sampling data could be obtained, this would be better, not for estimating the average but for increasing our assessment of the dispersion.

Additional data could sometimes be patched together from similar historical data, theoretical models, or looking at proxies, which in this case could be the returns from one of the other few US equity indexes. There is a popular "law of large numbers" in probability theory, which states that as the sample size increases, the sample average would equal the expected average from the underlying model. So if this is not the case, then one is best advised to reinvestigate their data and model, in order to reconcile a likely data quality issue.

Moving on from the central tendency (first-order) and dispersion (second-order) ideas, we will now focus now on higher-order moments, and think through them using our same 12-month simulated sample. It is worth noting that it is increasingly difficult to usually visualize beyond a couple of orders, or a few dimensions. But these high-order moments are highly important in risk modeling.

As we have noted already in the book, looking at higher-order uncertainty is generally an ignored task until it forcibly rears its face, such as in the market crash of 1987, the Flash Crash of 2010, the market melt-up of U.S. stocks, the crash of precious metals in 2012, the steep fall in Japan's stocks in 2013, or perhaps the large reversal in U.S. Treasuries later the same year. All of these events caught people by surprise, but this is a sample of many "surprises" that all appear the same in hindsight, and therefore should have always been able to be modeled to include them.

The terminology for moments is: expected "#-variation" moment. Where "#" is the number of the moment (e.g., 1st, 2nd, 3rd, 4th, etc.). And "variation" is either "raw" or "central". The average is the 1st raw moment, as we don't transform the data (hence it is raw), but rather just calculate the average.

The variance, on the other hand, is the 2nd central moment. Since we take the difference between each value and the "center", and then take the "2nd" power (square) of this value.

Of course standard deviation is $\sqrt{}$variance, or 3.8%. It is a trivial rule of thumb that the standard deviation will always be less than ½ the variance, whenever the variance is 4 (400%) or more. The equations in Formula G show the logic of when the square root of value is less than the value itself. We can also take this time to further connect some of the Chapter 1 distributions' parameters since we are on the topic of lower-order moments: Average of distribution=Variance of distribution$_{Poisson}$>Variance of distribution$_{binomial}$.

Now let's look at even higher-order moments, starting with the third central moment, standardized by (standard deviation)$^3$. This is a concept known as skew. Asymmetry, or the lower/left tail not being the mirror image of the higher/right tail, can not be detected by an average or the variance computation.

So instead, we look to weigh the larger deviations and also consider their directional sign. We see in this pattern below how the 3rd power retains the sign of the value:
$-2^1=-2, -2^2=4, -2^3=\underline{-8}, -2^4=16$.

And of course, similar to what we see in the standard deviation calculation, there is greater weight in those values most deviant from the center.

Note that since the straight units of deviations$^3$ is $\%^3$, here we try to scale this back so that the results are a little more standardized among moment calculations. The general approach here is to divide the result by $\sigma^3$, so that the $\%^3$ units easily cancel, and so that we can think of the 3rd moment results relative to the standard deviation. This also means that we now that the "standardized by ($\sigma^3$)" added to our moments terminology.

For our 12-month simulation, this third moment computes to -10.2%. So this very slight negative skew reflects an absolute central deviation of the 1st quartile that is greater than the central deviation of the 3rd quartile, by just 10.2% of the 4.3% standard deviation.

In financial markets, options valuation pricing includes many variables with Greek names that we'll explore in the next chapter, such as delta, gamma, vega, and rho. Gamma is the one that relates to the higher-order moment of considering the asymmetry in the rate of change of an option relative to the underlying asset price change.[xiii] Just as many choose to conveniently under-focus on (if not outright ignore) non-normal return assumptions in their pricing models, many do the same for the gamma term inside the derivative valuation equations.

The expected fourth central moment, standardized by $\sigma^4$, is similar to the concept of kurtosis. And $\sigma^4$=variance$^2$. Kurtosis differentiates itself from the variance idea in that kurtosis shows the amount of weight at the furthest deviations from the center, relative to what we get from the variance calculation. And fancy statistical parlance ensues for this concept, with platykurtic (-kurtosis) and leptokurtic (+kurtosis), to describe relative thin-tailed and fat-tailed distributions. Recall as well how we were discussing, without the vocabulary we now showed here, the tails in the Poisson hurricane example earlier in this book.

There is a different body of math for higher-order moments applied to the uniform distribution, as opposed to the normal distribution. Let's look at an example of the uniform distribution, which for now we can estimate with the change returns, on any purchase being equally likely to be any value between $0.00 and $0.99. This estimates the unit uniform, which can take on any value between 0, and 1.

We name a random variable $X$ to represent this change for any customer. And if we collect a large number $n$ of such random samples, then we begin to create a series of random values: $x_1$, $x_2$, $x_3$, ... $x_n$.

For a sample less than 50, say for die sums or monthly returns, we can compute higher-order moments with ease. But for infinitely large $n$, we need to rely on a convergence theory approach in order to guide ourselves.

Let's start by defining a concept called coefficient of variation (CV), which shows the amount variation relative to the average value. This formula below applies for any probability distribution that has a quantifiable variance (e.g., no data near $\pm\infty$):

$$CV_X = \sigma_X / average_X$$
$$(CV_X)^2 = variance_X / (average_X)^2$$

We now have a straightforward rearrangement of the bottom equation for $CV_X^2$, in Formula H. For the unit uniform, the average of $X^2$ happens to be 1/3, which is a fact everyone should commit to memory. Now $(CV_X)^2$ is $(1/3)/(1/2)^2-1=1/3$. Or $CV_X \approx 0.58$.

44

Instead of solving for the $CV_X$ using the generic equation above, for the uniform distribution we can Formula I. There we show, without proof, the CV for a uniform distribution that has a lower bound of anything, not specifically 0. And it can be applied for a stretched uniform distribution, not one that must be of unit length. Stretching all values of a distribution by a factor $n$, is the same as stretching the typical deviation[2] by $n^2$ (resulting in $n^2$*variance). So applying the unit uniform parameter of ($a$=0, $b$=1) into the broader uniform equation, we also confirm $CV_X=1/\sqrt{3}*1/1\approx0.58$.

In the next example, we look at the number of children in U.S. household families, and we will use a visual approach to understand the moments of this distribution. The data set has natural Poisson properties, such as parents being limited to having zero or a small integer number of children (e.g., no negatives, nor fractions). For this reason it is superior to the normal, or binomial models.

See the dataset, on Figure 3.5 below, from different U.S. government sources. We use the abbreviation "HH" for households.

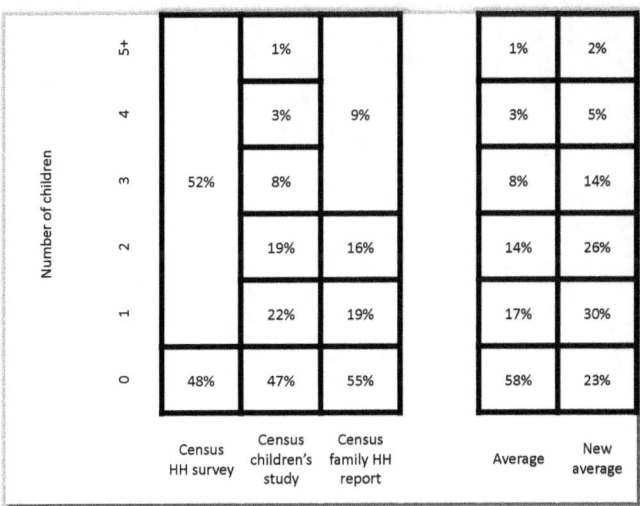

**Figure 3.5**

The first data column shows the empirical partition between household nonfamilies, household families with no children, and household families with children.[xiv] We removed the household nonfamilies set. The second data column shows the distribution of siblings count per family.[xv] We added a component at the column bottom, for childless families. And finally in the third data column, we show the number of children in household families.[xvi] Each of the three initial data columns have strengths and weaknesses, which is similar to how many data sets exist in the real world. The best analysts would again use as much of the sample data as possible.

We perform Poisson maximum likelihood adjustments (these techniques are beyond the level of this book) to account for censored data on cases of five or more children, and then consolidate our data columns into one master set on the right side. We can also apply a Bayesian likelihood adjustment to the number of with no children, considering intent and accounting for children not living with either parent. This adjustment allows for a reasonable estimate for no children and makes a Poisson illustration possible. We have a new average now in $1/5$, to $1/4$, probability range.

Now put the pencils and calculators down, as we will take a conceptual approach to the solution, starting with the Figure 3.6 presentation of the data. The straight line is a vertical representation (e.g., see horizontal grid and secondary y-axis) of the number of children.

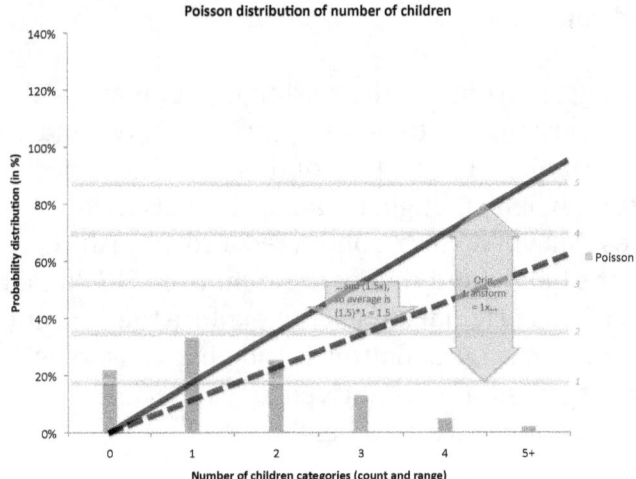

**Figure 3.6**

We think of the average as the sum of each probability bar times this straight line (child count). Which is equal to average*[sum of probability*(1)]. We can now think about a "penalty" parameter, $\lambda$, such that the lower penalized dashed line is assumed to be equal to sum of probability*(child count/$\lambda$)=1. This $\lambda$ would also be greater than 1 to lower the straight line to the dashed line.

The right balance for $\lambda$ is about 3/2, so that the probability sum, across child count categories, would come to 1. So we have just visually argued why the average is equal to the parameter $\lambda$, summarizing these equalities:

*$\lambda$*[sum of prob.*(child count/$\lambda$)]* = *average*[sum of prob.*(1)]*
*$\lambda$*[1, as assumed in earlier text]* = *average*[by definition, in text above, is 1]*

For the second moment, this is slightly trickier but still workable. We'll discuss the logic here. The key lies in keeping track not of the (child count) as we did above, but keep track of the (child count)$^2$. Variance is equal to [average($X^2$)–(average$_x$)$^2$]. In Formula J we show the average of $X^2$ equation, where the main point is that we instead solve for the expected value of 1-(child count).

We'll now switch gears from our discussion of moments, and discuss the characteristics of statistics associated between two paired random variables. The history of this topic dates back to the 19$^{th}$ century, when an Englishman Sir Francis Galton, cousin to Charles Darwin no less, conducted a relationship between a parent's height and that of their children. We'll take an abbreviation of this original data set to explore the concept of correlation, and covariance. Galton's study happens to be where the term "regression" was derived, in expressing "regression towards mediocrity" of heights.

Correlation simply implies that there is some dependent, linear relationship between two variables. And each variable value has a corresponding paired variable value, which it is being compared with. It could be a 1-to-1 link between parents and their children.

Dependency doesn't imply correlation, if not linear. For example, the weight of a book could equal the square root of the book's length. Certainly these two variables are paired and dependent, but not linearly so. And we mustn't confuse "correlation" and "causality". Tall children can be associated with taller parents, but tall children do not "cause" their parents to be tall. Causation models involve large sample sizes and generally a mathematical impulse, all of which uses approaches beyond the level of this book.

For the Galton's study, though not true generally for regressions, shorter parents (independent variable) would have short children (dependent variable), and the converse for taller parents. But these short children would generally not be shorter than their parents. And similarly the tall children would generally not be taller than their parents. If this were not true, then we'd currently be living in a world full of midgets and colossal giants.

Sometimes transformations are done to the variables to make the relationship linear, for example taking the log of economic data. Though log transformation can also change the correlation relationship. Afterall the positive outliers are being de-emphasized in either or both the variables. Other times, we can decompose a linear relationship into a string of different relationships, for example the fuel mileage as a ratio of a vehicle's age, changing when the vehicle is under age 10 versus over age 10.

See this quick example below for how the log transform can change the correlation.

$x = \{1, 10, 100\}$     $\rightarrow logx = \{0, 1, 2\}$
$y = \{1, 10, 25\}$     $\rightarrow logy = \{0, 1, 1.4\}$

$\rho_{x,y}$ is 0.96, but none here of these are 0.96:
$\rho_{logx,y}=0.99$, $\rho_{x,logy}=0.77$, $\rho_{logx,logy}=0.97$.

Regressions also are good if both the independent variable and dependent variable are binormally distributed. This avoids the issue of heteroskedacity (i.e., changing variances across the range of independent variable values) and (multi-)colinearity, both of which we'll later discuss in great depth.

Let's further explore correlations with a small visualization. Let's look below, at Figures 3.7, and guess which of the figures has the highest correlation?

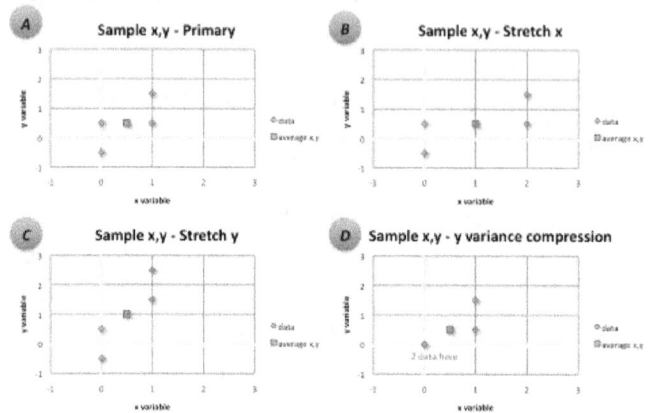

**Figures 3.7**

Usually in the real world, when someone simply states "correlation" in the context of a multiple probability variables, then they are usually implying the Pearson calculation. However, there are other types of correlation, namely the Spearman, and Kendall, which are both a little more nuanced. We'll start with the Pearson correlation formula:
$\rho_{X,Y}$=covariance$_{X,Y}/(\sigma_X^* \sigma_Y)$.

The basic idea of this formula is that we can find the expected variation of $Y$ (e.g., dependent variable) times the variation in $X$ (e.g., independent variable). This is the covariance term in the earlier formula. But to standardize the covariance into a measure confined between -1, and 1, we divide by the $\sigma$ of $X$, and the $\sigma$ of $Y$. On an aside, we can sometimes correlate a time series variable with itself (e.g., lagged one term) using similar mathematical formulae.[xvii]

So given the layout of the charts, it is clear that the vertical pairs of charts forces a highest correlation to be the one that jointly has the tightest-fit line through the data and also having the highest slope.  There is a slight greater weight on the former, versus the latter.

Chart 3.7C dominates Charts 3.7A and 3.7B on both measures.  The $R^2$ ($\rho^2$) fit for Chart 3.7C is also the highest, with a $\rho$ coming to 0.9.  Now Chart 3.7D seems attractive in shape, though the data cloud is not tilted as high in its angle.  So it's the next best, though has a smidgen weaker $\rho$.

Figure 3.8 shows the current data on the government's SBA 7(a) loans.  While it seems like a candidate for regression analysis, it is not appropriate for two reasons.  The first is there is there is large dependency between the variables.  Gross approval=(number of loans)*(average loan amount).  The second reason is that there is heteroscedasticity in the data, across number of SBA loans, shown in the change in data cluster lengths from the lower-left of the chart versus the upper-right of the chart.  It is both surprising and unfortunate how common it is for leading economists on Wall Street and in Washington, simultaneously violate both of these prerequisites, in order to rush out a geared point of view.

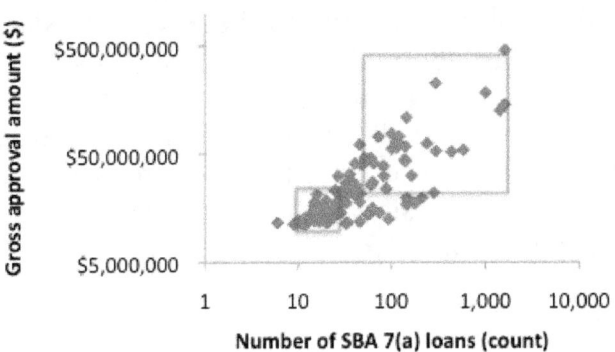

**Figure 3.8**

Now to show how the Spearman correlation is different from the Pearson, instead of the *X* and *Y* values, we look at how those values rank within the other values from that variable. Formula K is the set-up to convert the *X* and *Y* variable to the non-parametric correlation statistic, in describing probabilistic events we might come across.

The Kendall correlation is also a nonparametric rank coefficient that dampens the impact of outliers, though the emphasis here is just on the general direction between all pairs of *X* and *Y* values, and those from other pairs. Given the large number of pairings and greater volatility in looking at the ranks of underlying values instead of the correlation of ranks, we see in Formula K2 that the Kendall $\tau$ tends to be closer to zero relative to its otherwise similar Spearman $\rho$.

We should note with these different correlations that there are also methods that can be used to create nonparametric covariance formulas.[xviii] Another important topic in understanding correlations among paired variables is to consider not the entire distribution, but only the pattern at the extremes of a distribution.

This is where a body of copula theory is becoming more important to think about the relationships between tail risks (i.e., the extremes of a distribution). While this happens, it is a new and advanced enough concept that most analysts do not consider modeling tail-risks at all for their fundamental work, just as we saw looking at high-order risks were somewhat invisible by investment professionals that didn't have this in their traditional asset allocation framework.[xix]

http://statisticalideas.blogspot.com/2013/11/tail-concordance.html

In layman's terms these connections matter in novel ways. Say that U.S. intelligence officers has designed an elevator-operated escape path under the White House. However the risk of an attack impacting the electricity feed to the elevator was not considered. Then we would be ignoring the risk that the underground bunker may not being reachable when needed due to a higher-than-normal risk of the elevator being inoperable at the same time as a White House attack. Understanding the unique relationship between low-probability tail events, or the "wrong way risk" from concordance, is highly useful.

## Chapter 4: Stochastics

We will use this chapter to discuss probability theory in the context of high frequency, stochastic problems. There are many real world applications driven off these derivative calculations, particularly when the distribution parameters change through time. We earlier saw that the sum of continuously rolled die produce different outcomes over time, but the probability distribution was still always the same one we saw in Figure 3.1. But examples where the probability distribution changes over time include: the portion of the female population who reach their centennial birthday, the number of farmers in Europe, and the migration of birds every winter.

These latter time-evolved processes are often too difficult to solve in closed-form (e.g., without the use of a computer), so simulations are often used. Two popular techniques to do this are: the binomial simulation, and the Monte Carlo simulation. We'll describe these two different approaches here.

It should be noted that for a stochastic analysis, we need a complete set of high-frequency data, and not simply a description of how the data changes over time (e.g., a summary of the intergenerational difference of an economic measure). Let's look at an example below, to see the movement of a major hurricane, spotted on U.S. satellite, forming 50 miles southeast of Houston. Figure 4.1 below shows this storm, 10-miles over Galveston.

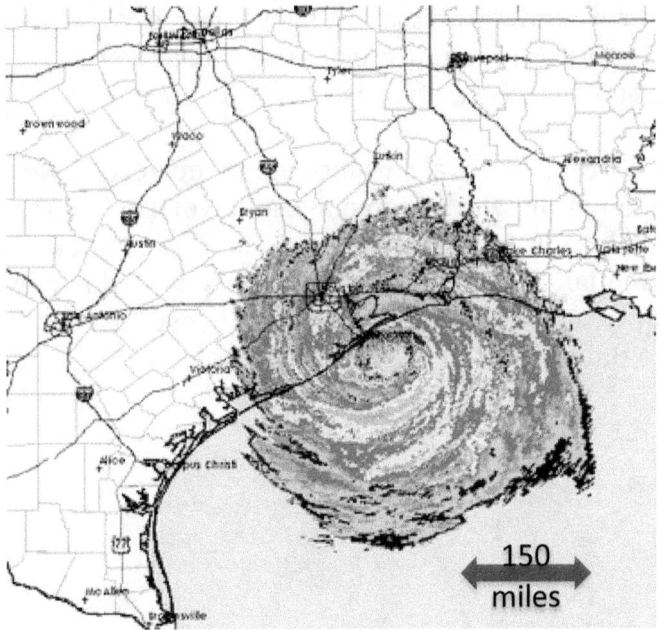

**Figure 4.1**

Ultimately we really need to know what's not what's up in the sky, but how this storm changes through time, as it descends towards earth. And what would be the probability distribution of its final impact. The changes in the storm can evolve in random ways (e.g., influences in wind, changes in speed due to weather or elevation, the width of the storm varying, etc.). So this makes simulation an easier method to model the storm's final distribution.

Let's say that emergency workers for the 2.1 million Houston residents want to know the probability of the storm center passing, in any direction, beyond a 50-mile radius of Galveston. [xx] First we'll discuss a hypothetical situation, with easy unrealistic, but easy-to-solve mathematical assumptions including no cross-winds. This can be thought of as an easier random-walk problem to, or further from, the storm's eye. For stochastics, the time unit would be instantaneously and immeasurably small, a case we have not seen so far in the book. But here with the set-up that we show, we'll explain how to interpret the storm's probabilistic distribution at any precise time, or any precise distance from the center.

With no cross-winds, and equal variance along the decent, we begin an approximation with a binomial framework in Figure 4.2 below. We'll lay out 16 simulated paths, or $2^4$ permutations for a 4-trial binomial. We could have scaled this up or down by a factor of 2, for twice as many or half the trials respectively, to better adjust to the number of trials we are comfortable working with.

What we see in the illustration below is that with the path's direction from the center, randomly either increasing or decreasing one per trial, the variance of any intervening distribution is simply scaled to be the trial number itself [recall variance is $n*p*(1-p)$]. Note that some math references use $m$ instead of $n$, but this changes nothing about the overall direction of analysis here. Also, this illustration below will be a handy reference later in the chapter.

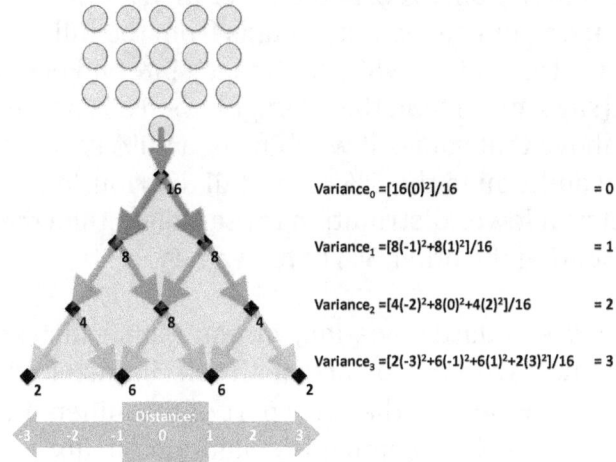

$$\text{Variance}_0 = [16(0)^2]/16 \qquad\qquad = 0$$

$$\text{Variance}_1 = [8(-1)^2 + 8(1)^2]/16 \qquad = 1$$

$$\text{Variance}_2 = [4(-2)^2 + 8(0)^2 + 4(2)^2]/16 \qquad = 2$$

$$\text{Variance}_3 = [2(-3)^2 + 6(-1)^2 + 6(1)^2 + 2(3)^2]/16 \quad = 3$$

**Figure 4.2**

The combinatoric formulae to solve the resulting binomial model is slightly complicated but workable, as shown in the Formula L set-up. The results could also be proven more formally, using an intricate mathematical theory that involves differentiating a unique "characteristic function". This unique function is the "moment generating function" (MGF); or the geometric expansion of which is a Taylor series expansion. We worked with the simplest variety of the MGFs in the previous chapter, when differentiating and then setting the MGF's $t=0$.

What is the main point of this mathematical proof? Our objective for stochastic analysis is to move from a discrete model and converge on a continuous model. So one that is limited to integer units of time, or distance, such as the binomial approximation. And from Figure 4.2, we can use the reference now that approximations of the normal distribution can linearly link the time elapsed, with the variance of the storm distribution.  In the process we swap out the concept of an integer number of **m** rows, and instead compute the variance along any precise time elapse, with the $\sigma$ being the $\sqrt{}$variance or also the $\sqrt{}$time.

Now is a good time to discuss how to think about CIs in the context of these simulations.  A CI is the inner distribution solved by subtracting unneeded critical tails from the full distribution.  If we think of the 95% confidence interval below a certain value (say **z** more than the average), then 5% of the distribution is above that value.  If we think of a 90% two-tailed CI, then in addition to the 5% upper tail distribution, there is symmetrical lower distribution (at say **z** less than the average) representing the other 5% of tail values.

Instead of the term standard deviation, we sometimes instead use the term "sigma" (with the lowercase Greek character $\sigma$).  As in operations Black Belts in the early part of the millennium, who were people that helped companies maintain quality controls to "six sigma" confidence.  We can use Figure 4.3 to look up the amount of a distribution that is below a set level (e.g., **z**-score).  Or we can follow the Excel commands shown in Formula L2.

| Z | x.x0 | x.x5 |
|---|------|------|
| 0.0 | 50.0% | 52.0% |
| 1.6 | 94.5% | 95.1% |
| 3.1 | 99.9% | >99.9% |

**Figure 4.3**

We use the backward approach to read the table, saw wanting to know the **z**-score for a 95% cumulative probability.  This level is found 1.6 along the vertical axis, plus 0.05 along the horizontal axis; the sum of which produces a $z$-score of 1.6+.005=1.65. Most popular statistical tables work similarly. Now just to test for comprehension, if we are looking to solve for the 95% symmetrical confidence interval, then we do not use a $z$-score of 1.65 for a cumulative distribution of 95%. This is because 1.65 is for a one-tail distribution of 5%, not 2.5% in each of the two-tails. Thus a cumulative distribution of 97.5% is instead needed (2.5%+95%).

Also keep in mind that the more complex the problem, the more the reliance on an open form (computer-aided) simulation to understand it. In the hurricane example this could be in the form of introducing heavy and random cross winds against the storm, as it descends. Or, there could be variations in the storm's movement that continue to change in random ways, as perhaps the storm interact with clouds during the tumultuous drop towards earth.

Before considering this though, we could take a look at another example of a binomial random-walk in the top illustration of this endnote (an example we'll use again in a later chapter on confidence intervals).[xxi] Or another example of a simple random time process is for popular gambling survival strategies.[xxii]

For more flexible and complicated examples, we can point to a government insurer to a country's pension system needs to model changes in employee behavior and investment asset returns, during future periods of the government fund. There would be a number of moving parts, particularly with the changes in the number of vested employees, which do not make a closed-form solution possible. Flight simulation models could also take advantage of this approach.[xxiii]

Right now as we emerge from this post-crisis recovery, many analysts attempt to surmise what the future economic period would look like. Will it be a low volatility regime, for example, or will things pick up gradually or in clustered spikes?

Invariably these pin-pointed questions don't take into account the full range of statistical complexity in the market behavior, but they are part of the learning curve that asset managers, and ultimately policy makers, seek to be on in order to take their work seriously. But it's these sorts of open-ended ideas that that can also make modeling the underlying characteristics in an open-form, Monte Carlo simulations, useful.[xxiv]

Let's lay out how these two modeling types differ:

| binomial | Monte Carlo |
|---|---|
| *The storm moves exactly one fixed distance (e.g., a mile), closer or further from Galveston, per unit of time.* | *The storm moves a random distance, closer or further from Galveston, per unit of time.* |

So as a general rule, we have the average ultimate outcome (e.g., storm distance) for the complete Monte Carlo simulation, which is greater than the average ultimate outcome for the binomial distribution. The additional convexity properties have generally –though not in every case- enhanced this final outcome, for the case of the Monte Carlo simulation.

The convexity theorem can easily be seen looking at average squared values for a uniform distribution model. One can skim over the detailed results below.[xxv]

*average of [function of X]    > function of [average of X]*

*If X is a unit uniform probability distribution, and the function is simply the square of the value X, then see the results below:*

*average $[X^2]$                              > $[average (X)]^2$*
*                                             > $[1/2]^2$*
*                                             > 1/4*

*And we noted that the average of the $X^2$'s for the unit uniform distribution was 1/3, so the convexity theorem holds, since 1/3 > 1/4.*

While the average may be higher, empirical results can often show a lower variance in a Monte Carlo approach, for two important reasons. The first reason is the misalignment of the variance in the Monte Carlo simulation versus the binomial distribution. Recall the trickiness in matching $\sigma$ to the interquartile range. While we set-up the formulaic expressions in Formula L3, one can also see Asha Kapadia's book for further background reading.[xxvi]

Now we can look at the second reason for the source of "variation" reduction when using the Monte Carlo approach. Recall that we have calibrated the variance of the binomial model, to match that of the Monte Carlo model. But this approach, using variance, would only align the lower-order risks (e.g., the second moments). But there are still higher order moments, such as skew or kurtosis, which would only apply to a small sample of non-normal distributions.

The Monte Carlo variations allows for a greater degree of variation outcomes per unit of time, unlike the binomial distribution model where the variance stays constant.

However the increased number of independent movements in this simulation can lead to greater mean reversion as well, versus the binomial distribution. We must appreciate that the variance by itself does not skew the direction of the storm's path; it just adds gyrations equally on either side of the more stable path trend.

The interactive patterns of the higher-order moments, which we have so far ignored in our simple storm example, will be used for more practical real-world stochastic formulas, as we go through later in the chapter. First let's look at what the full range of results would look like, based off of a small number of simulation results. Also we must bear in mind when looking at this, that the binomial simulation and the Monte Carlo simulation we showed provide different results. Similar to the issues we had in parameter estimation, we can have different, compelling statistical results derived from the same underlying data.

To explore the full range from a small sample, we will use a statistical technique known as a "bootstrap".[xxvii] The term bootstrap is attributed to a story where the main character pulls himself out of a swamp, by his bootstraps.

Bootstrapping allows one to further simulate a range of outcomes, from a small sample of collected data. Essentially allowing us to "bootstrap" to a better understanding of our distribution, particularly given the fickle volatility of small sample data. Imagine if one were bootstrapping with a data set that was: {0, 1, 1, 10}. One can appreciate that in this case that we need to better understand the upper-tail of the underlying distribution, given the one outlier from the small sample.

Due to similarities within permutations, we don't need to look at every bootstrap in all cases. We can save resources by only examining one permutation of any ordering (e.g., a combination).[xxviii] One can see Formula M for additional details.

We next calculate the mean square error (MSE) associated with the bootstraps, and for both the binomial and the Monte Carlo analysis. To see how the MSE on a small sample, we apply it for the variable $X$ below. Note the $X$ average is 4.7, already solved for in Formula M, and the $X^2$ average is 6.7.

|  | Bootstrap 1 | Bootstrap 2 | Bootstrap 3 |
|---|---|---|---|
| $X_1$ | 4 | 4 | N/A |
| $X_2$ | 8 | N/A | 8 |
| $X_3$ | N/A | 2 | 2 |
| average | 6 | 3 | 5 |
| square error | $(6 - 4.7)^2$ | $(3 - 4.7)^2$ | $(5 - 4.7)^2$ |
| MSE for average | = average of square errors immediately above | | |
| | | | |
| variance | $\frac{1}{2}(2^2 + 2^2)$ | $\frac{1}{2}(1^2 + 1^2)$ | $\frac{1}{2}(3^2 + 3^2)$ |
| square error | $(4 - 6.2)^2$ | $(1 - 6.2)^2$ | $(9 - 6.2)^2$ |
| MSE for variance | = average of square errors immediately above | | |

Through the bootstrap process we conclude that the "true" average and variance is now somewhere in the midst of these bootstrapped samples. For little extra resources to get here, we have a richer understanding of the true distribution as opposed to from the original small sample. And noted earlier in the Monte Carlo discussion of variance, without further information, we do not know what the true variance is. We see below, without proof, the general mathematical relationship of how parameters are biased from reality, regardless of simulation approach:

*MSE* $= (bias)^2 + true\ variance$[xxix]
*MSE - (bias)$^2$* $= true\ variance$
*MSE* $> true\ variance$

In our example, we show the MSE for variance of 13.3, while a true variance of {2,4,8} of 6.2. So the bias$^2$ we have versus the correct variance is the difference: 13.3-6.2.

From our binomial model simulations we generally have a negative bias in the MSE for average. This means we typically have a less-than-true computed value for average.  This is true from the simulation, and the bootstraps.  Given the bias differences between the two simulation methods, we see the critical point here is that the Monte Carlo simulation provides a closer reflection of the true distribution, though it may be improved by shifting those results inward (toward the center) somewhat.

Stochastic calculus allows us to, just from these arithmetic processes, solve higher level problems.  We want to derive relationships not for a linear output of a continuous process, but for many different functions of the continuous process. Staying with the storm example, Formula N has the set-up for the problem types we will now explore.[xxx]

The essence of the Formula N equations is that the "something" term, multiplied by the change in time, is brought in after solving down to the second derivative of $S$ with respect to $z$. We use the random variable $S$ to represent the final impact distance from the hurricane's start.  Convoluting these two variables (e.g., $z$ and $t$) is analogous to understanding and dissecting the additional terms of Taylor series approximation. What we learn in the latter is that some additional nuances of the trajectory's arc become essential for an increase in the underlying data value.  Financial technical analysis also subtly incorporates this information in connecting time, to the price support and resistance values that most market spectators focus on.[xxxi]  In the shaded area below, we solve for the "something" in the stochastic equation.  All that's needed is to see the terms move from the $\Delta$ in $z$ to the $\Delta$ in $t$ (the delta's "$\Delta$" symbol means small change).

| | |
|---|---|
| "something" | $\approx \frac{1}{2} * dS_{zz} *$ change in ($\Delta$ in z) |
| | $\approx \frac{1}{2} * dS_{zz} * \Delta$ in time |
| | |
| change in S | $\approx (dS_t + \frac{1}{2} * dS_{zz}) * \Delta$ in time $+ (dS_z) * \Delta$ in z |

Let's say that we are an insurer that wants to model liabilities in relation to the square of the storm's size, dampened by the speed of change in the hurricane. As an example, we can look at the following applicable model: $10*S^2-5*z(t)+5$.

The $S$ now carries additional weight due to the squaring of it. And with the stochastic calculus here, we can quickly solve this problem in a way that would not have been possible with the arithmetic process in the binomial and Monte Carlo techniques, solving for just the final storm distance $S$. The solution involves initially having to insert the appropriate values for the $S'$ derivative, into the equations from earlier, as shown in Formula O.

## Chapter 5: Samples, and errors

In this chapter we focus on a different topic in statistics. We take a deep dive into the concepts of sample size, confidence intervals, and hypothesis testing "errors".

Let's start with a common question we hear behind the scenes of many analyses: "what sample size do we need?" Any answer other than "the sample size for what?" is wrong. This is because we simply do not have enough information to complete this partial question, even to the basic magnitude.

The one thing that thematically will be repeated, and therefore will be clear to the reader by the end of this chapter, is that is having an even larger representative sample size (from what one currently has) is usually a good idea for better understanding a random variable. This is true if we are looking raw values, or even statistics of these values such as the average. In some odd cases, it is outright difficult to capture a sample to begin with (e.g., the target or cause of mass deaths in a war-strife region).[xxxii]

We can not simply walk around, as many professionals do, with enigmatic rules-of-thumb numbers for what a minimal sample size should be, without regard even to the type of problem they have. Yet this is often seen in informal scratch work during important meetings, and invariably we discover the uttered result is some round number such as 30, 100, or 1000.

In Figure 5.1 below, we present the results of a random number generator. Starting with a sample size of just five. What we show is the portion of data that is within one, and within two, standard deviations of the average. We repeat this process for a sample size of 10, 15, etc. And we stop the process at a sample of 40.

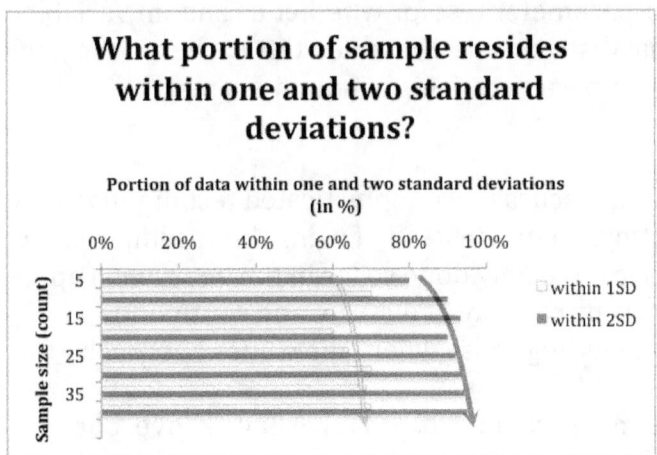

**Figure 5.1**

With a larger sample size, we generally see a rising portion of data within one (hollowed grey) and two (black filled) standard deviations. Put differently, there less tail distribution outside these symmetrical confidence intervals, as our sample size increases.

These portions also converge on specific values, as the sample size rises beyond a count of 30, at about 70% for 1 standard deviation; and about 95% for 2 standard deviations. And usually these values converge from below as the traditional calculations for variance from a sample under-estimates the true dispersion (also we'll soon see we don't need to have every data so there is flexibility), and hence we don't get enough of the distribution that we need. On the other hand, the average formula of a sample is an unbiased estimator of the population average.

The last two sentences of the previous page are some of the most important in this book, and they are concepts often glossed over in classrooms and beyond. This is the only time in the book that we recommend re-reading something now. Please carefully do so again.

Back to Figure 5.1, these use of two statistics form a non-traditional two-parameter test for whether a random variable is from a normal distribution. We noted this earlier in the book as another way of parameterizing a mound-shaped distribution.

In Chapter 9 we'll discuss more sophisticated techniques for distribution fitting. There, instead of using the specific *z*-score approach that we have only done so far in the book, we'll apply techniques to an entire array of discrete and continuous distributions. Including those where the values approach $\pm\infty$.

One can see a similar example to what we have above, but instead using the national SAT college admission exams, anchored on an average exam result of 500.

http://statisticalideas.blogspot.com/2013/07/standardized-test-scores.html

Just to reinforce the concepts here, smaller sample sizes do not bias our estimate of the true average. But they increases our uncertainty of the true variance (or $\sigma$ instead of variance). There are two joint methods to un-bias a small sample, and reduce the extra weight shown in the outer tails.

The first method is to increase the *z*-score level (the critical level) so that a higher score can realign to and scoop up a greater portion of the small sample, at a given $\sigma$ value. The second method is to enlarge the sample $\sigma$ such that it aligns better to the critical level established for a full sample. By "enhancing" the standard deviation on the chart we enjoy the same effect as "scrunching" the data, so that more of it fits within the statistical range for smaller sample sizes.[xxxiii]

We reconcile this sampling bias by enlarging to an "$n/(n-1)$" multiple versus the population variance to take into account the relative impact of consuming one degree-of-freedom in order to calculate the sample mean from a sample. Degree-of-freedom simply states that given a sample average, how many free values does one need to solve so that the rest of the sample alongside it comes to that average?

Now for a generous sample size of 30, this multiple is $30/(30-1)=103\%$. So we would need to increase the sample variance by 3%. This increase in the variance is equivalent to a 2% ($\sqrt{103\%}$) increase on the sample standard deviation. We then also swap out the $z$-score and instead apply a new statistic: the $t$-score.

We will now go through a more sophisticated example now to show cases where a larger sample size is necessary. Say that a restaurant owner is expanding into a new geographic region and wants to open a restaurant serving all-you-can-eat brunch, during the weekdays. The owner wants to capitalize on the busy lunchtime crowd, yet has no initial data to base what business traffic would be like in this new region. His of her objective is to provide just enough food to meet the needs of restaurant customers, while not over-preparing food that would then spoil.

To get an early feel for how much food to prepare, before opening day, the owner starts by visually surveying potential customers at the location of the new restaurant. He or she then notes the passerbys' estimated weight. The owner feels confident in his or her weight-guessing skills and begins to complete a pad of sample data, as shown in Figure 5.2, below.

| Sample number count | Weight (in pounds) | Food consumption (in pounds) |
|---|---|---|
| 1 | 169 | 0.69 |
| 2 | 168 | 0.68 |
| 3 | 183 | 0.83 |
| ... | ... | ... |

**Figure 5.2**

The owner believes that there is a linear relationship between one's weight and their food consumption (**FC**). And the owner feels that all the passerbys' weights happen to be equally distributed between 100 pounds, and 200 pounds. And their **FC** is 0.1 pounds for every 10 pounds of weight, above 100. So a unit uniform for the **FC** could be used here, with equal probability of values between 0 and 1 (a unit).

The statistics task at hand is to create an **FC** estimate, and know that the sample **FC** from the notepad is within a certain accuracy of the true **FC** when the restaurant opens. Right now the owner has a sample size of just three on the notepad, with an estimated **FC** of 0.73 pounds consumed per guest. This was derived as the average of {0.69,0.68,0.83}.

But there are at least two significant things wrong with this initial approach, and this is precisely where strategies related to crude reality begins to step in the way of book theory. Given the small sample size, one just doesn't have a good sense of how much statistical-weight to give any specific sample value within the {0.69,0.68,0.83} values. Perhaps 1/3 of the customers will consume about 0.83 pounds, or maybe really 3/4 will over the long haul. We simply don't have a good enough sense with this small odd-numbered sample. The risk from undersampling 0.83, would imply a lower estimate of the true *FC*, which could leave the crowds with a food shortage. The science of small samples tends to treat them as greatly vague representations of a large sample, when in fact we have just seen that this may not at all be true.

The second thing wrong with this initial approach is that a small sample size of three isn't broad enough to give a sense of the entire range of possible *FC* possibilities. We saw a hint of this in the bootstrap procedure of the previous chapter. There could be other *FC* estimates other than {0.69,0.68,0.83}, which are yet to be included when the sample size grows.

Let's say the owner looks now at a fourth and fifth *FC* estimate of {0.55,0.19} and enters that next on the notepad of Figure 5.2. Both of these values are below the range of the first three *FC*s of {0.69,0.68,0.83}. And the new *FC* estimate now falls from 0.73 pounds, to 0.59 pounds.

Can we just borrow from the previous example in this chapter and increase the sample size from 5, to 30? The answer is no, and for multiple reasons. The first reason we reject the sample size notion of 30 in this case if that we are not testing a simple statistics characteristic from a normal distribution. And the second reason is that we are more interested in our confidence on *FC* estimate and not the dispersion of the *FC* estimate itself. For these specific reasons, a larger sample wouldn't just be a nice-to-have, but it would be needed to gain more confidence in our estimate.

But as anyone will tell you, real world resources are finite and so we can't simply collect an infinitely large sample. Decisions have to be made (and well in advance), on how the test should be calibrated. Say the owner decides to simply ask in this case, what sample size is needed, to be 90% confident of being within 5% of the true average.

We can review the closed-form solution to this in Formula P, and we'll show soon how to visualize this with a graphic. In Formula P, we start with an assumption that the sample $FC_{variance}$ and the true $FC_{variance}$ are equal (the subscript is simply short-hand to help identify the variable statistic). This is many times not true, but if we feel that the final sample size will be in excess of 30, then this assumption will correctly save a lot of time. Also, since our ultimate test concerns the $FC_{average}$ and not the $FC_{variance}$, this assumption will not have a material impact on our analysis.

So we wrap up the final part of the closed form solution, started in Formula P. And we pay close to the **bold** line of the equations below, as this equality will be named the "ultimate equation" and used throughout this chapter as it is the workhorse for many other related sampling problems across different distributions.

$$(Coefficient\ of\ variation_{TFC})^2 \leq [5\%\,/\,y_p]^2 * sample\ size$$
$$\textbf{(Coefficient of variation}_{TFC}\textbf{)}^2 * [y_p\,/\,5\%]^2 \leq \textbf{sample size}$$

To solve this ultimate equation, we showed earlier in the book that the unit uniform has a variance of $1/12$, and an average of $1/2$. We can also use Figure 4.3 to help understand the $y_p$ level desired. This figure can work in this case, because the sample size will ultimately be large enough that the $t$-value would collapse into the $z$-values for Figure 4.2.

We now know the 90% confidence in this case will comes from a 95% cumulative distribution. And the appropriate $z$-score is then 1.645. Our solution is a minimal sample size is $(1/12)/(½)^2*[1.645/5\%]^2$, or 361.

Now let's see visually how the choice of sample size with the restaurant owner would impact how wide his or her 90%

confidence interval would be, for being within a ±5% sampling error. The 5% band on a ½-pound $FC_{average}$ is 5%*½=0.025 pounds. Beyond a sample size of mid-300s, we see in the figure that the 90% confidence width comes below this 5% error band. And at least 90% of the time, the actual simulated $FC_{average}$ (the dark line shows one simulation below) is in fact within this 90% confidence range.

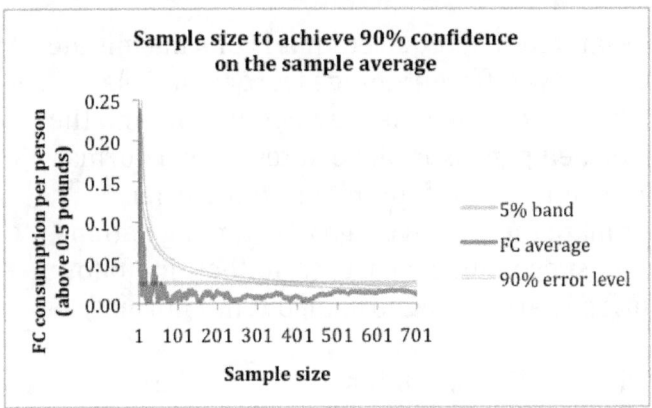

**Figure 5.3**

The initial sample values of {0.69,0.68,0.83,0.55,0.19} have a $FC_{average}$ of 0.59, which is 0.09 more than the 0.5 pound true average and more than the 0.025 we earlier showed as our 5% sample error band.

Even without this precise calculation above, the main point here is that as the sample size increases, the volatility in this sample $FC_{average}$ decreases. It also stays close to the 90% theoretical error level, which exponentially falls below the 5% error threshold once the sample size is greater than 361.

So now, can the restaurant owner or anyone else suggest that if they want to estimate any statistic within 5% of its true value, 90% of the time, a sample size of at least 361 would do? Of course not. Else the popular Gallup polling organization would be able to whip out U.S. Presidential election predictions with similar accuracy parameters, by only screening 361 potential voters. Meanwhile, in practice, thousands of voters are screened for many national polls.

One can note that so far we had not needed to know the size of the underlying population, in order to know the minimal sample. The minimal sample size is independent of the population size (other than it must be less than it), and it not a continuous function of the population size.

We could have a sample size required that is of a magnitude larger than 361, if a few different things were taking place. The first way is the degree to which the true population, and the particular statistic being measured, deviates from a normal distribution. For example, we couldn't use our earlier procedure to estimate the true average GDP among nations, since similar to most economic measures, it does not follow a normal distribution (perhaps lognormal, but not normal)

This non-normalcy condition, combined with the large fluctuations in the population per country variable, would cause us to see extreme variances in the true GDP per nation. And thus we'd have a more complicated variance and $CV^2$ to contend with in our ultimate equation. Hence a larger minimal sample size than 361.

The second way that the minimal sample size could be much larger than 361 is if instead of wanting to estimate the average from the sample, we want to estimate a more complicated statistic with a precise confidence interval. If for example, we wanted to precisely estimate the aggregate daily revenue of the restaurant, instead of simply the average FC.

Aggregate revenue implies that we need to take a guess at both the average consumption value, and number of customers. This also leads to a more complicated $CV^2$ portion of our ultimate equation, by replacing $(1+CV^2)$, for $(CV^2)$. For log-normal distributions, which are distributions whose logged values revert to a normal distribution, such a change in $CV^2$ would be $(1+CV^2)=(e^{\text{standard deviation}})^2$.

And the third way to have a significantly larger sample size is if we change the model so that each sample is not an identically distributed representation of the underlying population. Say instead of wanting to look at the American approval rating for their president, we instead wanted to look at the American approval rating of their governor where they live. The approval rating of a Californian for their governor has less to do with the approval rating of a Floridian or a New Yorker to their respective governors. So a larger sample size would need to be aggregated, versus if we were only measuring citizen's views towards the same president.

An astute reader will notice that we are still generally sampling for confidence on the average statistic, and not on the dispersion or other characteristics, a topic which is beyond the scope of this book. Now let's go through an example where we look at a topic involving both sample sizes, and confidence intervals.

Since we know confidence intervals vary based on where it is in the distribution, it is important that any seemingly extreme views are appropriately expressed. For example, when working as a department head at the PBGC, we would create actuarial reports on the probability of insolvency at a range of future dates. The public would naturally zero in when these probabilities are very high (>90%), and perhaps become too laxed when the probabilities are very low (<10%). In other words, probabilities and confidence were given far greater weight then they deserved and hence the communication "messaging" can help to hedge this. In reverse, one can undue biases in the messaging to understand the underlying information an organization likely has or how the media would typically present this information.

Another example is at the start of 2012, when the Federal Reserve Board began a new process of announcing their own forecasts for the bank's key discount rate. Wall Street and the media eagerly scrutinized this first new report, but then analyzed it in simplistic ways (e.g., average initial rate-increase vote among Federal Reserve officials, or average path over time, etc). This is the exact similar basic way they crunched the comprehensive GDP benchmark revisions making them less valuable then they could have been.[xxxiv]

A lot of statistical value was left undisclosed from the Federal Reserve's projections. For example, to this day few have broached the topic of "confidence interval" information that is inherent in the information they present. And if we were to delve deeper into the Federal Reserve's own forecast uncertainty, then we could garner more interesting details surrounding the future path from the near zero-interest rate policy (ZIRP) in effect since December 2008.

For example, until early 2012, the Federal Reserve had repeatedly suggested that economic conditions warrant "exceptionally low levels for the federal funds rate at least through mid-2013." But such a comment is not precise since we don't know how "economic conditions" is defined, or what "exceptionally low" means, or what would occur after "at least mid-2013".

Looking at Figure 5.10 below, we see the actual January 2012 Federal Open Market Committee (FOMC) reported projections.[xxxv] To illustrate the idea of variance of views within the FOMC, they note: "the shaded bars represent the number of FOMC participants who project that the initial increase in the target federal funds rate (from its current range of 0 to ¼ percent) would appropriately occur in the specified calendar year."

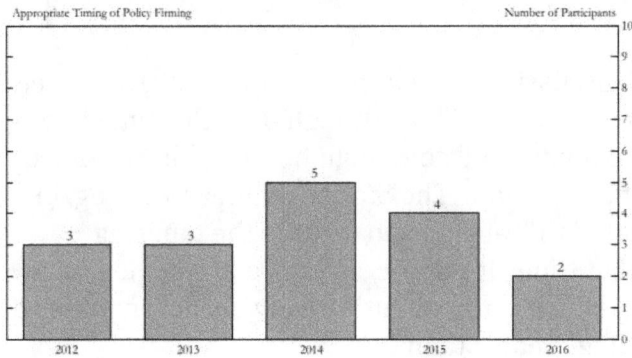

**Figure 5.10**

As we have done previously, we can sum the individual probabilities alongside the time axis, creating a cumulative distribution across time. With 17 mutually exclusive participants, each count is modeled with 1/17 probability. In actuality there may not be 17 voting officers at any given meeting. And we know, though can't model at all, the influence one participant (Chairman Bernanke) has on the other participants. For example, current Chair Yellen as the Vice Chairman almost always votes the same way then Chairman Bernanke did, so this would alter the probability distribution if we take that information to imply causality.

In any event, the main topic here is we are looking at this problem assuming identical and independent weighting of participants for this report. And this report is a new heightened degree of FOMC transparency, since the financial crisis, even if we can not ascertain the people behind the votes in Figure 5.10.

So at the start of 2012, how would we have calculated the probability that we would have policy firming by the end of 2013? We can sum the counts per year, right off of the illustration:

| | |
|---|---|
| *p(timing of policy firming ≤ 2013)* | = *N(events ≤ 2013) / N(all events)* |
| | = *6/17, or ≈ 35%* |

With the FOMC discussion during the autumn of 2013, we see this probability is now less likely, though over this time it was certainly in the ballpark and sovereign bond yields bottomed in the middle of this period. The Statistical Ideas web log, in June 2013, updated this analysis to include the tapering discussions at that time. It put a 1/5 chance of rate firming by the end of 2014, which is not quantitatively unreasonable and must be something greater than 0.[xxxvi]

Now we ask a more interesting question: how much <u>confidence</u> is there that this 35% estimate represents the true probability? Remember that this is as of early 2012. Without knowing the confidence interval, we are bound to be overly confidence in the 35% probability and then over-hedge the chance of rate firming.

This is a typical error that investors make and would instead better off if they were more humble in considering that they are often unaware of the correct answer despite their own statistics "facts". It would also prevent many from being surprised, let alone "misled", by what they feel were strongly confident probabilities that the FOMC would begin to tighten in a specific way in 2013.

So there are a number of reasons why this 35% estimate lacks precision versus the true probability. One reason is that we have a very small sample size, which is an issue at time that we've already discussed. Another reason is that there is rotation in representation among the FOMC staff, in terms of who gets to participate in these projections and meetings. So this would impact voting decisions in a way unrepresented from just the projection data. There are also four more reasons, three of which are: the staff are not identical in their insights (heterogeneity), the staff may have biases in their perception of the all incoming data (e.g., we have the descriptive terms for this such as "hawk" or "dove"), and as we already noted some staff may be greatly influenced by specific other participants.

The sixth reason, which is one of the most important for market observers, is that policy officials can hedge their personal views in the market, providing a signal opposite to how they vote, in order for falsely lead the market in a positive direction. This statistics wrinkle, in part, led to a massive "surprise" in the September 2013 near-unanimous decision to not taper the historic bond buying program, when the market investors (not policy wonks nor strategists) had expected that it would have to start in order for rates to rise by 2015.

These variables below start us in computing a variance of this 35% chance of a rate hike by the end of 2013. To use a pun given the topic, we'll describe the approach here as a binomial probability model, with a "twist":

| | |
|---|---|
| $p$ | = probability of timing of policy firming $\leq 2013$ |
| $q$ | = probability of timing of policy firming $> 2013$ |
| $n$ | = sample size |

Now we first compute the variance of the number of participants projecting policy firming before the end of 2013, and then show the logical leap on how to convert this variance for the cumulative probability of participants. We can see Formula P, and we'll use the "$\approx$" symbol since we still don't know with certainty the cumulative probability, we are just estimating it from the sample.[xxxvii]

With our variance equation in place (e.g., the **bolded** expressions in Formula Q), we can simply plug in the results given to us in the FOMC projection:

$$variance_{cumulative\ probability} \approx 6 * 11 / 17^3$$
$$\approx 35\% * 65\% / 17$$
$$\approx 0.0134$$

So the $\sigma$ of the estimate is $\sqrt{variance}$, or 12% in this case. Therefore a symmetrical one $\sigma$ interval on the probability estimate of rate firming, by the end of 2013, is 35%±12%.

One should be able to see now, in a way that hadn't been true before, that the error of one's estimate is relatively large here. Hence we are working with a fairly imprecise estimate of the true value. Put differently, a 1 $\sigma$ confidence means that about 1/3 of the time the actual result would be outside of this 35%±12% range. The fact that Chair Yellen in a March 2014 responded to a question that rate firming may occur about six months after tapering ended (which is likely to end in 2014). It is possible that we in fact were in one of those outer 1/3 probability events.

Also while this may not initially seem intuitive, it doesn't matter that there is a distribution bulge shown at 2014, or what the rest of the distribution post-2013 looks like! We are only interested here in the probability, immediately surrounding the 2013 year demarcation.

With this variance we can now return to our trusty ultimate equation to solve for the minimal sample size needed if we wanted to generate the 90% confidence of being within 5% of our underlying statistic. And we see that the sample size we need is not only far greater than 361, but impractical within the use of FOMC participant data (again, we don't have anywhere near that many projections from which to study).

$$(Coefficient\ of\ variation_{TFC})^2 * [y_p / 5\%]^2 \leq sample\ size$$
$$(n - n_x) / n_x * [y_p / 5\%]^2 \leq sample\ size$$
$$(17 - 6) / 6 * [1.645 / 5\%]^2 \leq sample\ size$$
$$1984 \leq sample\ size$$

With this symmetrical confidence we should see that the information "noise" is far greater then the underlying "signal", and it is this noise that certainly needs to be understood. Another issue with this set-up is that our estimate was for a symmetrical CI, which is only applicable when the estimate is mostly in the middle of the range. But this would not make sense if our estimate is generally at the ends of the distribution, where we show an asymmetrical CI is expected:

*Each tick marker on this range of the unit uniform distribution is 10%:*

```
|-----|-----|-----|-----|-----|-----|-----|-----|-----|-----|
<-|------->                                        ← (A) ends forces CI asymm.
             <-----|----->                         ← (B) middle so CI symm.
                        <-------|-> ← (C) ends forces CI asymm.
```

For example, if we had instead looked at the two $\sigma$ confidence about the probability of rates hiking before the end of 2012, then we'd end up with an impossible probability of less than 0% on the low end of the range: 18%±2*9%. Seeing the (A) above, it makes sense therefore to alter our CI to look at a distribution that shifts inwards slightly, in order to completely fit more comfortably between 0%, and 100%.

In probability notation, we use the expression **F** for the cumulative probability distribution (the most common form), and *S* for the survival distribution (useful for actuarial math). The survival probability is the opposite of the cumulative probability, so *F*+*S*=100%, or *S*=1-*F*. We use this in Formula R to create an exponential formula that is bounded by arithmetically symmetrical CI that we seek. In doing this we also reverse the time calendar to start backwards from the 2016 year, and end with the 2012 and 2013 projection years.

We can see both the symmetrical and asymmetrical CIs in Figure 5.11 below. Since the probability estimate for rate firming by the end of 2013 is 35%, this more central value implies the CIs will be similar though distinguishable. It is also not often discussed among even quantitatively-geared individuals, when the probability is over two dimensions (rate hike, and year), then there is another CI for the timing year as well. This is shown on the horizontal range of the rectangles below.

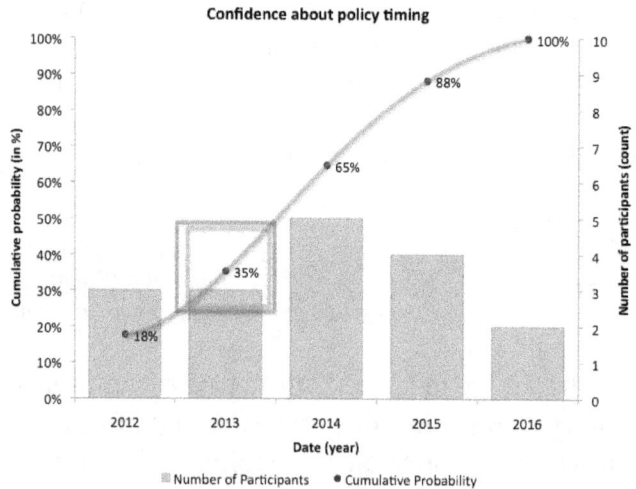

**Figure 5.11**

Now let's switch gears and discuss the meaning of Type 1 and Type 2 errors. These are the errors we generally associate with hypothesis testing on random values. We'll look at an example of a gold mine. We'll see through the example the connection between CI, and the critical level that needs to be established for statistical tests.

Let's get started with a visual of a square shape field. On this land we are offered to purchase a land parcel, which may contain gold. This plot of land is shown as a box, with a "?" on it, in Figure 5.12 below. The parcel is 15 miles away from the suspected gold trove center, which is the lighter colored square inside the figure.

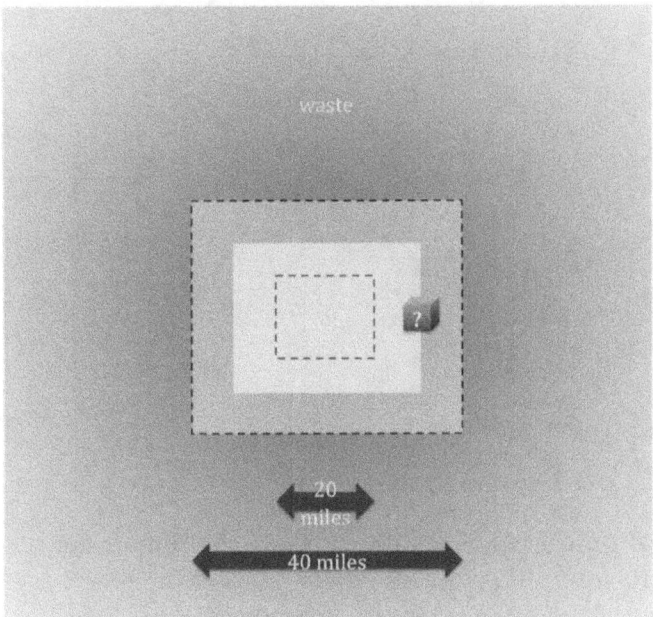

**Figure 5.12**

Metal and other commodity fields have hidden value distributed below the surface. And only probability models, created from tiny probe data, can estimate the potential value of a space before a full drilling or excavation operation.

A method to assess the quality value of the ground below the surface, is to use traditional hypothesis testing. In Figure 5.13 below we start with the knowledge that 90% of the gold is located in the 20-mile length squared center. We label this gold portion as "**p-gold** = 90%". This implies that there is a 90% chance of gold being found at that location, not that 90% of the ground is gold enriched. Put differently, there are 360 miles³ of gold parcels under the 400 miles² land.

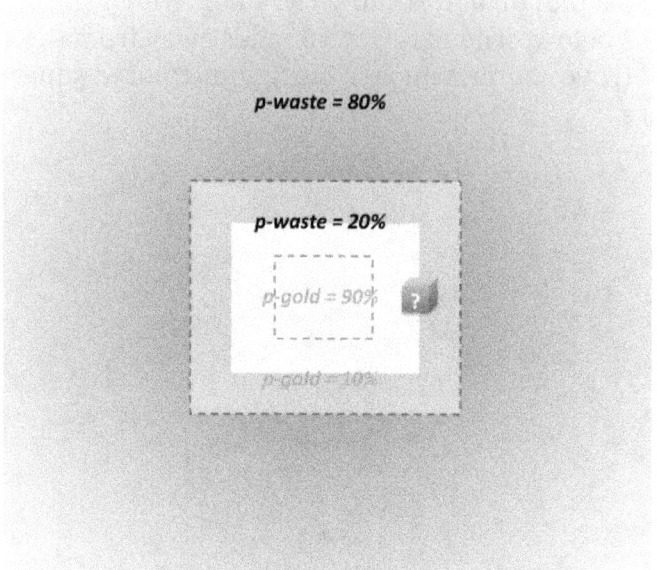

**Figure 5.13**

Outside of the **p-gold** area of the illustration, the remainder of a 40-mile length squared region is where 10% of the gold is and where 20% of the pure waste is. So let's calibrate our understanding of the problem. Would it be more unlikely to find gold:

○ on the inner 20-mile square trove center, or
○ in the rest of the 40-mile square perimeter of this inner square?

It should intuitive that we are less likely to find gold further from the center, where only 10% of the gold is, and where 20% of the waste is. But the offer is for a land parcel between the two areas above, located 15 miles from the trove center. If only accept land closer to the center, then there is still no guarantee that we would have gold there. Equally bad, we would be rejecting many parcels that have gold at a further distance from the center. And because we have information on alternate composition probabilities (i.e., waste) on the ground other than gold, we have the information to solve for the error probabilities of both conclusions.

The essence of the Type 1 error is the probability of not accepting something true. For our gold field, if our criteria for exploring a parcel of land is set so far from the trove center, then surely the probability of gold being found will be small. We know, for example, that beyond a 20-mile length square it is only 10%.

The question can therefore be, why not just simply set this critical distance as far from the center as possible? The answer to that is while the Type 1 error would then reduce, the second error (Type 2) would increase. The Type 2 error is the probability of accepting something as true, when it is in fact false. For example, the probability of taking a land parcel assuming it's not waste, when in fact it is waste. This type of error, which can be high if one lowers their Type 1 error, is not often considered in the real world and would take alternate distribution in order to pursue.

Similar to a few sort-cuts in science[xxxviii], the field is incorrectly doubling down on p-value emphasis by creating a Type 3 error[xxxix], which I am not fond of. They suggest one can be right for the wrong reasons, when the alternate rationale is a wrong one is still wrong but for the wrong reasons. Right for the wrong reasons is rejecting the null but the alternate is wrong. This is not a (Type 3) error however.

In our example we have provided this alternate hypothesis that a small parcel of land not enriched with gold, would be waste, even in a vast gold field. Returning to our land parcel, the probability of assuming it won't be waste when it in fact is, is equal to 20%.

In hypothesis testing of this sort, we are challenged to find the right balance in analysis so that we minimize both error types. But there are utility weights on both these errors as we hope to minimize them. For example, a clinical screen for the HIV virus could be set to such a high a threshold that there is a minimal chance for a "false positive" that falsely scares the patient. While a low threshold could mean that many infected would go undetected. While both are important, clearly the practical utility here of the Type 1 error (a false scare) is far less critical than the utility of the Type 2 error (not detecting the virus).

What we've seen in this chapter is the important to understand the tolerance about a sampling threshold, and how much dispersion there is in errors about a probability estimate. We also know that there are utility considerations to consider as well, in addition to a need to minimize errors in analytical conclusions.

## Chapter 6: Advanced confidence intervals

In this chapter we will take an extended look at calculating CIs, including within new settings such as regressions, and causal analysis. Let's start with an exotic hypothetical example of a kidnapping that has taken place in an Indonesian public park. Say that we are scientists deployed to Merdeka Square, inside Jakarta. Merdeka Square has many venues and traffic arteries, and at 0.4 miles$^2$, it is 1/3 the size of New York's Central Park. See Figure 6.1 below for a layout.

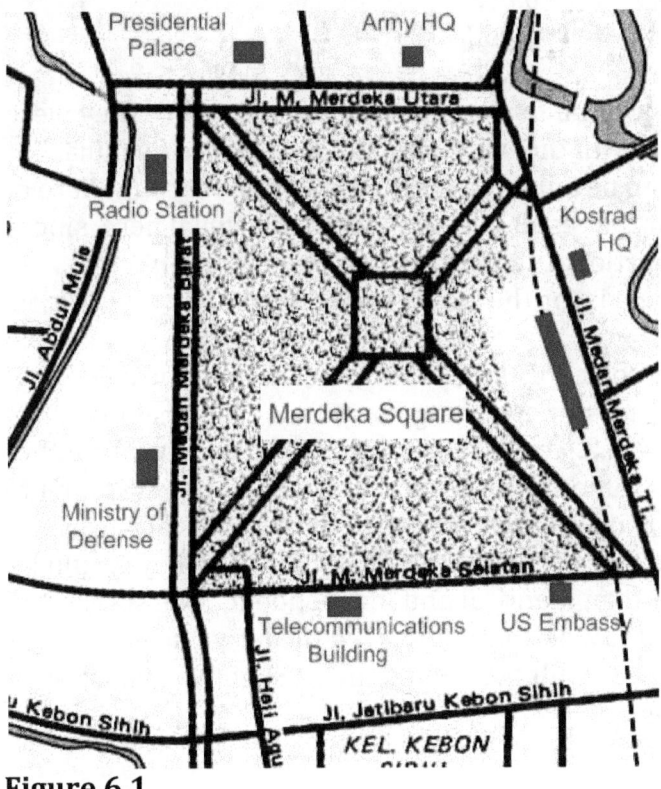

**Figure 6.1**

There are seven eyewitnesses that have given a sense of where the kidnapping has taken place. But these locations are different despite each one claiming to have the same faith in their ability to recall where the location was. So to start we impute a similar CI, around the locations each witnesses identified.

Of course Merdeka Square, technically, is not really a square shape but more of a trapezoid. This fact is neither here nor there. We'll model it still with a vertical (latitudinal) and horizontal (longitudinal) axis emanating from the southwest corner of the park. We can then decompose the horizontal and vertical differences, in each of the witnesses' locations, and show that each direction's variances are close to one another as shown below.

$variance_{longitude}$ = 0.003
$variance_{latitude}$ = 0.002

There is a slight wrinkle about these variances, since they were both computed with the basic population variance formula. They have yet to be adjusted for the small sample count here of witnesses. But the result will still be roughly equal here, since the same proportional statistical adjustment of $n/(n-1)$ applies. Where $n$ is for the seven eyewitnesses.

So we can now solve for the 90% CI in relation to the eye witness' location estimate, using Formula S. The result is shown in Figure 6.2. And again we only need to solve for either longitude or latitude since the directional deviation was about the same. We can also pool these two directional variances together, since there is no reason to assume that they would be anything more than identical and independent. One statistic would also lead to a better result given the larger size of information. The result here is a $\sigma$ that is slightly smaller, at over 0.05 miles.

The Pythagorean Theorem (e.g., $x^2+y^2=z^2$) provides a radial distance $z$, given the latitudinal ($y$) and longitudinal ($x$) variances. For this purpose, think of the radial distance as the shortest distance between two points. So the average distance of 0.08 miles comes to a CI of $\pm 0.15$ miles. We illustrate these eyewitness locations and the approximate 90% circular CI, respectively, in Figure 6.2 below.

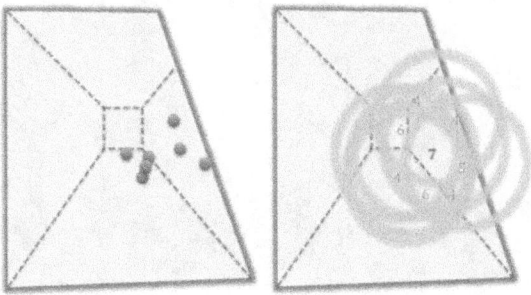

**Figure 6.2**

This is similar to "triangulating" with satellite as they have recently for the MH370 flight, except here we think of the full circle range as 90% CI it contains the true kidnapping location, 90% of the time. In Figure 6.2 we show the number of witnesses including in our CI for a sample of regions. The "7" value implies the region was included in all seven witness' CIs.

One might be wondering, if the CI calculations for two independent dimensions (e.g., latitude and longitude) were to combine into one circular CI, would a squaring adjustment needs to apply. In our case, if we instead had a 90% confidence in both directions, then by independence doesn't this imply an 81% (90%*90%) circular confidence? The answer is no, and we'll show why starting with the visual set-up in Figure 6.3 below.[xl]

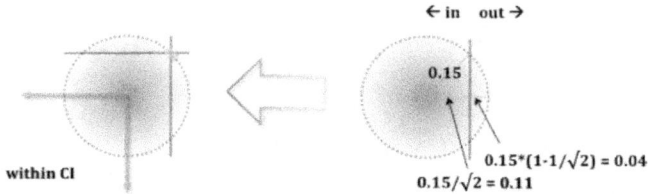

**Figure 6.3**

What Figure 6.3 shows is that we can have one partition of the circular CI along either the latitude or longitude. The results are symmetric either way. The amount that area that comes from both side of the longitudinal divider some to 0.1 miles (~0.05*~1.8). We can then derive, in Formula T, a probability model appending the relative weights from both sides of the partitions.

Here is a good time for us to discuss the relevant concept of marginal probability densities. The left part of the illustration above shows the idea of a marginal density. It is the amount of the total probability slice located at the infinitesimally small chord, cross-sectioning the distribution. While this density value can be more than 1, the value of this density rate summed over any small width must still be in a probability range between 0, and 1. One can see Formula U for working with marginal densities and scaling factors, depending on the size of how spread out probabilities over some space.

An example of this is that an equal probability of a given day of the week is different that an equal probability of a given day of the month. This difference is a scaling factor.

We tie this analogy together by using p(in) and p(out) is to confirm that the conical three-dimensional probability to approximate the weight we have from the bi-normal distribution in Figure 6.3. To solve for this we sum the approximate geometric shapes from the cone atop a cylindrical base, and use the marginal density information. This is shown in Formula U2.

The result from this visual approach above is that the 90% confidence applies for two-dimensional circular intervals. A limited short-cut developed is to take a handicap of 1/confidence range$^{[1/(3-\text{number of }\sigma)]}$, which for about 1.6 means we need to (100/92) multiplied by the 81%. Or back to about 90%.

What we now know is that about 6 (90%*7) of the circular CIs should generally contain the true kidnapping location. Based on this information, we can think of a composite of all regions labeled "6" or "7" for where the true kidnapping location is.

Bear in mind that the geometric center of the confidence intervals is likely to be in the region labeled "7". Though it doesn't have to be, particularly as the true region can be in areas "6" as well, and not limited of course to a "7" region. Recall from the previous chapter that a Type 2 error would be important to keep in mind if we were to increase the circular CIs (not size of the circle, but size of the confidence level) from 90%, to 99%. This is the error of thinking of areas distant from the "7" region as likely to contain the kidnapping location, when it really shouldn't at all be included as an area of interest.

There are two ways that we can, in the real world, increase the precision of our CI. The first again is to increase the sample size, which among other things reduces the sampling variance we've discussed a number of times so far in the book. The other way is to blend in "diversified" new data. For example, asking for additional attributes other than kidnapping location so that we can better match the kidnapping location. For example, we can ask what nearby landmark was near the kidnapping location, in addition to where it occurred.

Through the sub-additivity probability theory briefly discussed in Chapter 1, we know that compounding two independent risks together creates less risk than the sum of the component risks. We can see this variance reduction from diversification, in Figure 6.4 below, where each random variable is uniformly distributed integers from 0, to 10.

| Sample #, and Statistic | Random1 | Random2 | Diversify via average of (Random1, Random2) |
|---|---|---|---|
| 1 | 1 | 5 | 3.0 |
| 2 | 2 | 1 | 1.5 |
| 3 | 10 | 9 | 9.5 |
| 4 | 6 | 2 | 4.0 |
| 5 | 5 | 7 | 4.5 |
| Variance | 11 | 9 | 7 |
| Average | 10 | | 7 |

**Figure 6.4**

We need to be clear that the CIs that we have do not indicate that there is a 90% probability the kidnapping took place where the witness stated. That is a wrong interpretation of probability in general. There is only one true kidnapping location, and either the kidnapping took place at a given location or it did not. There is no "probability" associated with it. To be precise, the location provided by all combinations of 6 witnesses' locations (or even their geometric centers in this case) can not all locate the exact kidnapping location because they all pointed to different locations.

Now let's explore CIs from a new example, here using linear relationships. In Figure 6.5 below, we show the average heights of Americans from 2003, to 2006. This was published in 2008, by the U.S. Department of Health and Human Services.[xli] We only look at the early years of 2, through 10, where the linear relationship holds up better. We also take the average of {male, female} heights, so only one cohort height is shown for a given age.

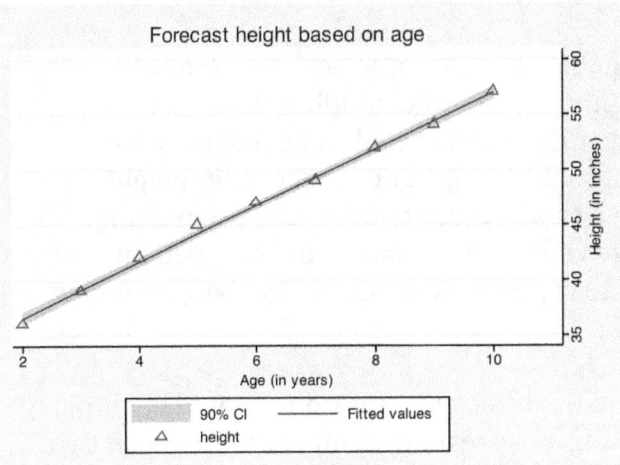

**Figure 6.5**

While we'll explore more dimensions in a dedicated regression chapter later, we'll discuss the linear fit aspect through CI here. This is important to give exposure to for linear regressions since they are often (mis)used in discussing data, in many aspects of society.

By using the average height, per age, we have removed extra dispersion called expected process variance (EPV) of the height values at each age. This height dispersion per age was a standard error of under 0.3" in each gender. But which since the genders themselves are typically 1" apart, there is a standard error of over 0.3" when we pool together all the heights from both genders.

The regression builds towards the correlation² can be done with the concept of the EPV. So for an upward trend of data, we first establish a variable $a$ to represent the distribution variance of the average height values shown in Figure 6.5, while we establish variable $v$ to represent the EPV. Both $a$ and $v$ of course must the greater than zero. The $\rho^2$ is then equal to $a/(v+a)$. From the regression we have an error decomposition that is from both the trend and the error about it, and in the same vein the total height variance here is equal to $a+v$.

But note again that for this example we have removed the EPV dispersion and only focus on the middle value of all the children's heights for a given age. We see that even this average trend line shown doesn't connect every height data, but we still have a small amount of deviation remaining. Why is that? And if we wanted to replace the illustrated line with others that seek to capture the thrust of the data, then how would it be done?

These are some of the basic questions one should be thinking about when looking at a presumably linear fit between two data. Sure it's ok to not connect all of the data on a perfect line, but there should be a good explanation for what's driving the deviations. Also since the majority of the data are generally in the middle of the distribution, the outer parts of the distribution get greater weight in determining the fit of the relationship. This is due to the fact that there are less sampled data in the outer regions. And also the deviations², which helps define variance, give greater weight therefore to the largest deviations.

A regression relationship suggests that the slope of Figure 6.5 is the estimated constant growth in height, for each increase in age. We can see Formula V, where we abbreviate standard error as "SE".

Now we can decompose the height and age relationship into a best fit line, consisting of both an initial height plus this height

94

growth expected per year of age. This means that that all the variability of the fit can be expressed through two sources of errors: the estimation of the level and of the slope (the slope is from the smaller sample generally at the ends of the data cluster in the regression).

Combined, these two errors form the convex CI region we see illustrated. The resulting CI shows to be thinner for the middle ages (e.g., age 5, through 7), while the confidence interval expands more at the more distant ages from the middle (e.g., ages 2, or 10).

Confidence in the estimation greatly increases when we estimate heights, from closer to the middle ages. One sometimes doesn't have a choice in where among the independent variable (horizontal axis) they must make an estimate, so it's more important that these relative CI regions be kept in mind.

We should re-emphasize that the 90% CI may appear more narrow relative to what one might expect from a data cloud, but we are not fitting all heights per age, but just a more stable average height per age. We have therefore eliminated much of the EPV as noted earlier. We reinforce this with the following calculations for comparison:

$$variance_{average\ height} = variance_{height} / (n)$$
$$= (standard\ error_{height})^2 / (n)$$
$$\approx (0.3)^2 / (about\ 250)$$
$$\approx 0.0004$$

$$SD_{average\ height} = \sqrt{(variance_{average\ height})}$$
$$\approx 0.02\ inches$$

How many of the average children heights should be within the 90% CI? This is an easy question. Of the 9 ages, 8 (90%*9) should be in the CI. And in Figure 6.5, we see that all but the 5-year olds average height is in the CI.

95

The final topic on CIs for this topic is for what the CI would "infer" is we wanted to predict the average height for ages we hadn't modeled. Say a 1-year old or an 11-year old. Since we are moving away from the defined CI region, this other estimation interval is named the "predictive confidence". Given the un-modeled age whose height we are predicting, we would have a larger predictive confidence versus the main fitted CI we see in the illustration.

We show in Formula V2 what is required to solve for the predictive confidence for an 11-year old. In order to solve for this value of 0.4", we need to first know the value of $S_{xx}$. Recall that the variance of $X$ is simply the sum of squares divided by the distribution count. So we can calculate this simply by multiplying $X$'s variance by the count, which we show as 60.

This might take a second-pass reading to understand, but it is well worth appreciating that the predictive interval, similar to the CI, is intended to be normally distributed. Since we have the slope CI mixed with the level CI. And a normal distribution mixed with a normal distribution is also a normal (the average of multiple normals we'll see is a Chi-square). We can see this the approximate normal mixture below, where we mix a 2-trial binomial with a 2-trial binomial to simulate this effect.

```
probability
50%   |       @
25%   |@      @      @
value -2     -1      0      1      2


probability
50%   |             *
25%   |      *      *      *
value -2     -1      0      1      2


probability
50%   |             #
25%   |      #      #      #
value -2     -1      0      1      2
```

We now mix these distributions with the 2-trial binomial weights of {¼,½,¼}, related to the distributions above, centered about {-1,0,1}. And our total result is:

```
probability
38%   |             #
31%   |             *
25%   |      *      *      #
19%   |      *      *      #
13%   |      @      *      *
6%    |@     @      @      *      #
value -2     -1      0      1      2
```

To see whether this matches up to the probabilities for a 4-trial binomial, we show those probabilities here using the same techniques we learned in Chapter 1. So $_4C_{(m+2)}(½)^4$ is:

*For m* $= -2, -1, 0, 1, 2,$
*p(m)* $= 1/16, 4/16, 6/16, 4/16, 1/16$

And we see it is a perfect fit to the illustration above! Now for the following chapter, we'll go into the topic of multi-dimensional visualization. This will be an interesting advance on some of the concepts we've discussed, in previous chapters, concerning intuitively thinking about and visually solving probability problems.

## Chapter 7: Visual solutions

In this chapter we will discuss Bayesian statistics, and we'll show how to visually implement allied ideas using a probability curve. Real life data are sometimes fluid, and we get to think about the impact of new data versus our pre-existing notions.

Think about touring a particular large home for the first time. One may get an initial rough sense of what the home looks like as they walk in the door, but a lot of this is from our imagination using previous home walk-through experiences. There is much intrigue about the many possibilities that remain inside this particular home. But as one begins to tour different rooms, they immediately get to switch over from their imagination, to the non-fictional reality of what they are observing. Soon, before even getting to see the final room, one has made a fairly good assessment of what the entire home is all and most of the intrigue has vanished.

This is how some of our statistical models work, where we have an underlying prior model (an initial observations or known facts) and ten we combine this with the new observed distribution (perhaps a continuous flow of new information). And while we initially weigh our prior understanding to form a posterior understanding, this soon evaporates and we quickly become dominated with our understanding from the new observations.

Here's another example of how this could work. Say that we are driving on an unknown highway without clearly posted speed limits. We may only have a vague sense of the local traffic laws based upon intuition. And this would be considered our "prior". Eventually we would see other cars either pass us by or get passed up by our vehicle. So these would be the new observations. We would be able to update our understanding of what is likely the true speed limit distribution from this combination of information, forming our "posterior" distribution.

Let's go through an example to help illustrate this, starting with just a small amount of hypothetical sample data. We will visually show a nontraditional "leaf" impression to help understand this data better. Say that professionals of an organization are evaluated on an annual review system, where a grade of "-1" is worst and a grade of "+1" is best. One professional is Bill, who after four annual reviews, has the following grades: {-1,0,+1,0}.

It's plain to see that Bill is averaging at a "0" performance, and this is our prior. After two additional years, two new grades were received: {+1,+1}. How could we interpret these new observations? Would we say that Bill is suddenly a "+1" performer?

That would be too sudden as it discards after two reviews, all of the four reviews that came before it (only one of which had Bill at "+1"). But we can't deny that getting two "+1" reviews is suddenly moving in a direction of clearly better performance. Look at the left illustration, in Figure 7.1. Here all six years of performance reviews are laid out, along the horizontal axis. The prior "0" performance is shown as a bright marker at the origin.

Now one annual review at a time, we show the impact of the new observed annual performance on our posterior understanding. This posterior distribution is scaled along the vertical dimension. In the middle illustration, we just show the fifth year's performance as a "1", both vertically and horizontally. All that remains are the four prior annual reviews, still along the horizontal axis.

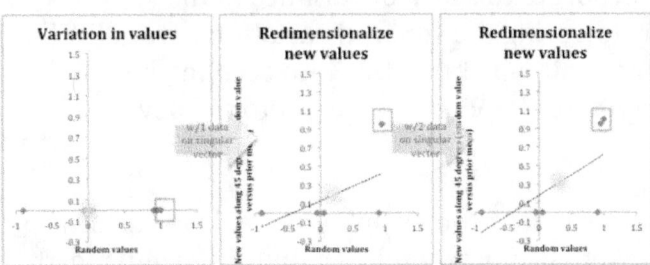

**Figure 7.1**

If all five performance data were given equal weight, and plotted vertically and not just horizontally, then we'd have a linear fit line at a 45° angle. It would then be easy to measure the posterior performance as the vertical distance of the bright marker along 45° line.

But here we only have one observed data that is given important weighting in updating our prior. Our linear fit line only moves up to less than 45° angle, and the posterior performance is therefore not as high up as it would be if it were on a higher 45° angle.

Next we look at the right illustration where we then observe year 6's "+1" performance as well. In a similar fashion, the best fit line gallops quickly towards just below 45°. Of course it couldn't rise beyond 45°, so there is a sort of ceiling at that level.

The posterior performance accordingly moves is now closer towards "+1". This major updating process makes the leaf geometry interesting. And this visual is how we can imagine

101

the updated posterior probabilities of otherwise complicated probability problems.

Note that there is a second aspect to this problem since the successive new reviews were consistently at "+1". So there is greater confidence associated with the regression line tilting towards those vertical data on the leaf illustration. We can secondarily incorporate this lack of variance in the observations by lowering the squared variations overall. The regression solution already takes this into account, by providing even greater importance weight to the new observations.

This type of example often comes up in the real world. In professional basketball, for example, an undrafted and mostly unknown player came onto the scene in early 2012 for the New York Knicks. A sudden high streak of great performance lead to a statistical phenomenon for Jeremy Lin, where there was an overweighting of his recent performance. He instantaneously became a national craze, named Linsanity.[xlii] We know now that the sudden star performance then wasn't to be suddenly extrapolated forever into the future. But for the moment, it was just Linsanity.

In actuarial science we use a term named "credibility" to mathematically assign an approximate portion of weight on new data (e.g., new grades or performance). Since we will use it a little further still, take a moment now to appreciate the basic set-up of the credibility formula below.

*credibility = sample size / [sample size + (v / a)]*

*Similar to in the previous chapter that "a" represents the variation of all new and old grades. And "v" represents the weighted variations of the new grades and the old grades.*

*For example, we noted above that the variation of the new grades is zero. And lower v values reduce the denominator (i.e., the bottom portion of the ratio). Combined with an increased sample size, we would certainly increase our credibility portion.*

Returning not to Bill's performance reviews, we can now incorporate the formula we just showed to see the evolution of our probability understanding of the performance. We start at the top with the four initial review, and then in the middle we incorporate only the value of the new "+1" performance reviews. In the bottom illustration, we show how the probability distribution changes further given the consistency of Bill's year 5 and year 6 grades of "+1".

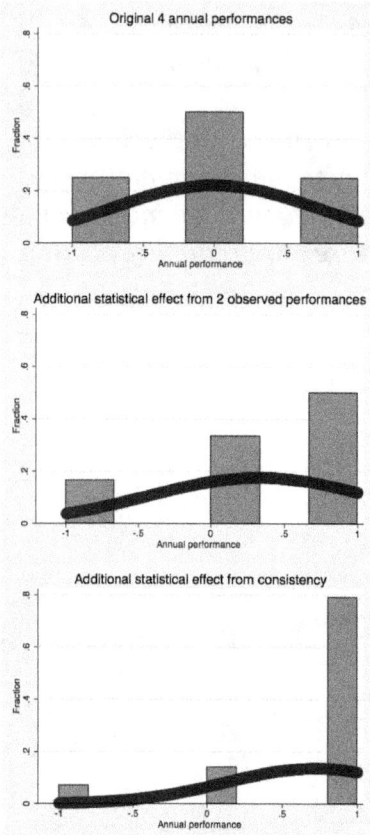

**Figure 7.2**

We can also see another example of credibility analysis, in thinking through how the FOMC can consider the weak July (and August) labor market data for the September 2013 meeting.

http://statisticalideas.blogspot.com/2013/08/credibility-of-labor-market.html

At this point in the chapter we want to further discuss the concept of an "outlier". This concept represents any data that does not fit into the main data trend. Unfortunately the term is instead incorrectly used to reference many things, including references to people in a derogatory manner. Across high-tech and low-tech industries, if we look hard enough, we can always come across stray data values for a number of reasons. We should be on the lookout for errant data and analyze what its purpose.

The most popular reasons are related to scaling issues, coding errors, small sample sizes, rounding errors, missing data, poor assumptions (particularly about the randomness of the data or completeness of what the data represents), outright data corruption, and data manipulation.

One example of an outlier was during 1976, when Nadia Elena Comăneci received a "1.00" score in the Olympics. The crowd was initially confused as she had given a brilliant performance. Though soon everyone understood on their own that there was a flaw in the Omega scoreboard, which was only manufactured to carry only the first three digits of a score. Amazing, right? See up until her performance, no one has ever received a perfect 10.00 score, so to save money only three digits were manufactured for the scoreboard. Meanwhile the crowd sensed that there was of course a scaling error, and that the third "0" in 10.00 was being truncated. And the false value of "1.00" was instead being given. This was an outlier in action. And the crowd soon went from confusion, into a bursting of cheers. But how do we know when such errors occur when things are not on display, but instead when a machine is processing a voluminous amount of data for a user? Many times the answer is not as well, and hence this is one of the lurking risks for Big Data activities pushed onto all industries by the technology sector, and which we all need to be mindful of.

Let' see if we can use outliers to help understand the strengths and limitations of data we hope to make direct conclusions from. The 2013 comprehensive benchmark revisions to the U.S. GDP[xliii] caused research to be done on the economic value of digital R&D[xliv], and intangibles. There was also research on how to connect these economic inputs to company financial metrics[xlv] and national growth accounts. To discuss outliers it will be helpful to look at the basic template for this analysis, but first with a hypothetical data set, which can lay the groundwork for larger studies on the effects of the GDP revisions.

See the five data in Figure 7.3 below. Now there are two ways we could draw a linear relationship through those data. One way aims to minimize the perpendicular distances between the data and the line. A second way aims to minimize the squared vertical distances between the data to the line. We see the orthogonal (solid line) and linear (dashed line), respectively, in the illustration below.

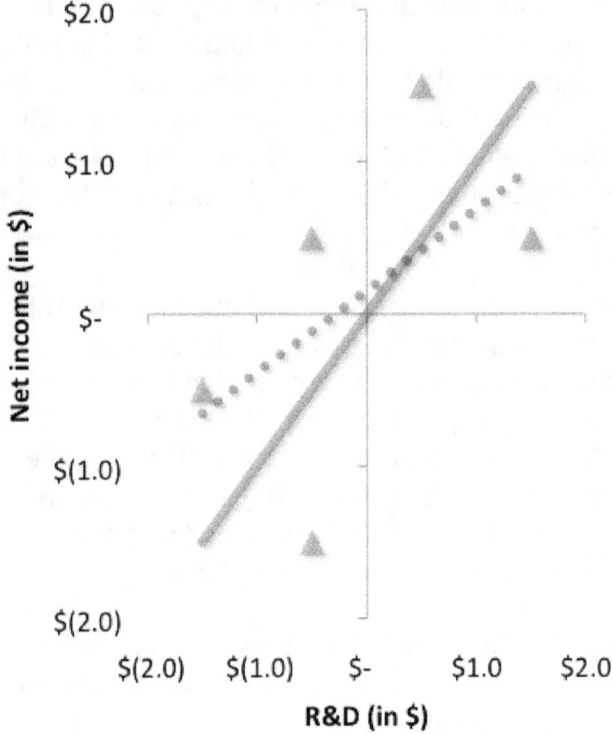

**Figure 7.3**

Surely these two lines would converge if we had a large cluster of tightly linear data. But if we start there with all of the real underlying GDP data and just allow a computer to blindly hijack the regression procedure, then we would forever fail to understand the broader importance of the instability between these two methods. This instability shown in this example above helps us understand the real "story" behind our data, which we also need to understand the outlier ideas of how to think about inevitable strayed data.

Just for edification, the linear sum of the distances$^2$ in the solid orthogonal line is 5, while in the dashed regression line it is 3.7 (this is the only measure where linear residuals are better under the latter approach we will therefore name least-squares). The reason for this is that the latter is able to gain ground on the outer data, which it more closely matches to. Using the linear regression approach, the data located at (-$0.5 R&D, -$1.5 net income) would by far be the main outlier among the five data. Using the orthogonal approach, the story is far less clear. Add additional outlier to the cluster and we'd add to our complications in automatically highlighting an outlier from the pack.

Let's nonetheless work out the regression concepts with this dataset. With some mathematical manipulation, we can rearrange the sum of distances$^2$ into a ratio that is $\rho^2$. In order to do this we first code the previous sum of distances$^2$ as "SSE", and the total vertical sum of squares as "SST". Of course the SST for both lines is the same. And starting from the (-$0.5 R&D, -$1.5 net income) data, and working clockwise, it is: $-1.5^2+(-0.5)^2+0.5^2+1.5^2+.5^2=5.25$.

The SST=SSE+SSR. So this SSR, or the model balance, is the sum of the expected regression line (similar to the variance of hypothetical means $a$ in the previous chapter). Which, after the line is vertically shifted to intercept the origin, is just equal to the slope coefficient $\beta_1$ multiplied by the independent variable $X$ value. We show the proof connecting SSR, to $\rho^2$, in Formula W.

For the dashed linear line and the solid orthogonal line, the correlation $\rho$s are 0.5 and 0.2, respectively, per Formula X. Something we learn from this process is that outlier selection is perilously connected to the choice between orthogonal, and least-squares relationships, are interconnected. Resultantly, we can't do what many people mistakenly do. And that is test whether a data is an "outlier" by simply seeing if the $\rho^2$ increases with this data's removal.

So far in this chapter, we've discussed a leaf-chart example, to incorporate new information. We discussed some topics concerning outliers via regression. In Figure 7.4 below, we examine the range of mixture and credibility relationships between two random variables. The statistics topics in these chapters have some overlap, as we saw the final row approximated by looking at binomial distribution mixtures, in the previous chapter.

| x-value | y-value given a x-value | | Continuous mixture | Bayesian | Bühlmann |
|---------|-------------------------|--|--------------------|----------|----------|
| gamma | Poisson | | negative binomial | gamma | gamma |
| inverse gamma | Exponential | | Pareto | inverse gamma | inverse gamma |
| beta | Binomial | | | beta | beta |
| normal | Normal | | normal | normal | normal |

**Figure 7.4**

Now is a good time to reassess the second column, and incorporate this table into observations that we might notice in the real world. We can then combine this with any relevant prior distribution in the first column. In doing this we can better understand the type of posterior distribution that is most likely to occur, given a particular $y$ mixed with $x$ (otherwise stated as $y|x$) distribution mixture from Figure 7.4. These posterior distributions are shown in the two rightmost columns.

We next apply one of these examples (e.g., the normal|normal combination) in looking at the Bayesian result of the U.S. 10-year Treasury yield, as of late 2011. The timing of the analysis is just after the Standard and Poor's downgrade of U.S. sovereign debt. We first log-transformation the pricing data before mixing.  This is fine since the convex yield distribution for short time periods is tightly distributed, above zero.  Borrowing from terms we earlier discussed, we therefore have a low CV in the discontinuous, economic periods data we will analyze.[xlvi]

It is important that we feel comfortable that the volatility is not so high, in a low rate environment, that the rate can somehow be shocked into negative territory.  This has never occurred on a nominal basis in the U.S.  For this modeling comfort to occur, we need to diminish the volatility accordingly as rates are at lower levels, and thus prevent this large downward shock.

The modeling adjustments are similar to those in what is named the Rendellman-Bartter interest rate models dynamics, which otherwise fall just less than the parameter constraints of what is named the Cox-Ingersoll-Ross model.[xlvii]  We pick this late July time to identify the start of market participants' exasperation with the contemporaneous lack of progress on the federal debt negotiations.  We see similar concerns reverberating a couple more times.  The most recent was two years later in late September 2103, and there it was to a smaller extent.

In the 2011 context, the economic concerns were sharpened with information that the U.S. GDP was cumulatively revised down 5%. Despite this and other shocks, we see an observed interest rate distribution that has moved lower a bit. Of course then, as well as in late 2013, we can expect a posterior distribution that should incorporate higher rates, from the observed levels.

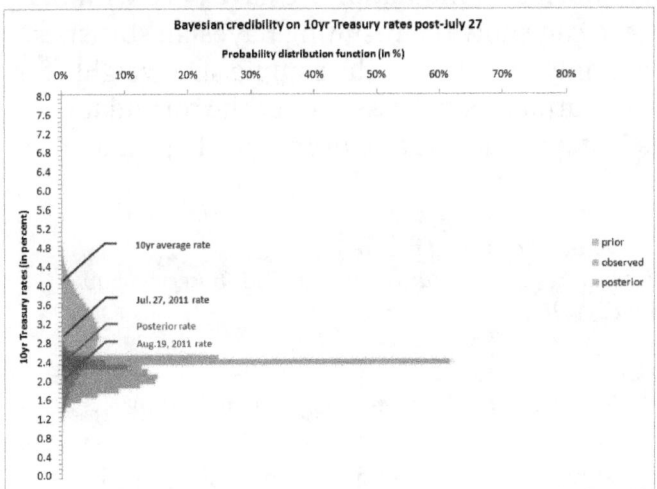

**Figure 7.5**

The newly observed interest rate frequency distribution is the bottom curve, and the prior interest rate frequency distribution is the top curve. And the posterior distribution is always the middle curve between the two.

A prior distribution, or "$\pi$", to some degree has an unknown average, since market stochastic data can always have some inherent uncertainty to them. This prior distribution would be modeled with the fundamental market data being normally distributed, or "$f(x|\lambda)$", about their longer-term trends. And the Bayesian posterior distribution derived from this treatment would also be a normal distribution, as we see the new probability curve morph as it did in Figure 7.2.

We don't show the formulas with the $\pi$ and $f(x|\lambda)$ terminology, though we see the basic point through the formulae over the next couple pages. In statistical parlance, we are connecting the posterior distribution model with the prior distribution model, with a mixing distribution that is named the conjugate prior.

Given the normal|normal combination we have here with U.S. Treasury rates, we can show the resulting Bayesian statistics. Right now we are only focusing on the sample size weights for the ultimate distribution. One can skip over the formulae below, which is just provided as a reference for Figure 6.4.

*avg. of posterior*
*= [(sum all observed values / observed pop. $\sigma^2$)*
*+ (avg.$_{prior}$/sample variance$_{prior}$)] / [(days post July 2011/observed pop. $\sigma^2$)*
*+ (1 / sample variance$_{prior}$)]*

*variance of posterior*
*= 1/[(days post July 2011 / observed pop. $\sigma^2$) + (1 / sample variance$_{prior}$)]*

In a final visual example will be used on the exponential distribution, which is often used in finance and science as a method to calculate continuously compounded growth rates. We also use the exponential distribution in actuarial mathematics to describe mortality rates.[xlviii] We can see Formula Y for equations, and note that mortality rates can be used to describe the value of "death" rates in many applications, such as life annuities, financial options and derivatives, and product failure warranties.[xlix]

Let's think about the variance of an exponential model. We'll first logarithmically transform the random variable and look at the variance of that transformed variable. We also have a random variable $X$ that has a unit uniform distribution, so it could be between 0%, and 100%. And the value of $X$ will govern the continuously compounded rate of our exponential distribution.

Random variable **Y** represents our financial account in one year, after starting today with a balance of one €. So $Y=e^X$. A 50% simple rate of return in a year yields 1*(1+50%)=€1.50. And a 50% continuously compounded rate of return instead yields $e^{50\%}$=€1.65.

We showed earlier in the book how to calculate the variance of **X**, which we'll anyway show again below. But at this point, the thrust of this example is to think about the variance of **Y**, ultimately with visual techniques. First, here is the variance of **X**:

*average X* $\quad = \frac{1}{2}$
*average X²* $\quad = 1/3$
*variance of X* $= $ *average X² - (average X)²*
$\qquad\qquad = 1/3 - (\frac{1}{2}^2)$
$\qquad\qquad = 1/3 - 1/4$
$\qquad\qquad = 4/12 - 3/12$
$\qquad\qquad = 1/12$

Let's now review derivative functions of **X**, which we will use to estimate the Taylor series formula for $Y=e^X$. Then we can solve the formula of a specific **X** value of about ½ (e.g., **x**=50%).

$f'(x_0)$ $\quad= $ *slope of f(x₀)*
$f''(x_0)$ $\quad= $ *acceleration of f(x₀)*
$f(x)$ $\quad= f(x_0) + f'(x_0)*(change\ in\ x\ vs.\ x_0) + f''(x_0)*(change\ in\ x\ vs.\ x_0)^2+...$
$Y$ $\quad\approx e^{\frac{1}{2}} + (e^{\frac{1}{2}}) * (x - \frac{1}{2}) + (e^{\frac{1}{2}}) * (x - \frac{1}{2})^2$
$\quad\approx (x + \frac{1}{2}) * e^{\frac{1}{2}}$

*Note that if x = 50% in the final equation, then Y is:*
$Y$ $\quad\approx (50\% + \frac{1}{2}) * e^{\frac{1}{2}}$
$\quad\approx (1) * e^{\frac{1}{2}}$
$\quad\approx 1.65,\ or\ same\ value\ that\ we've\ solved\ for\ above$

The Taylor series is named after an early 18th century mathematician of the same name, who was able to approximate a power series with discrete sums. The Taylor series shows up repeatedly in a number of probability applications. For background, we can get a quick calculus review setting in a James Stewart textbook.[l] On an aside, the natural log base **e**, or powers of the same base, can of course be expressed as a power series.[li]

In a number of applications, the vast majority of the series value is locked up in the initial terms, which most people find intuitive since they are the lower dimension terms. So the series approximation of **Y** results in a precise 1.65, when the **X** is ½, as shown in the formula above. But the higher powers (the terms with taken to the second and third powers, and beyond) start needing to become more important as we move further from our initial **X** baseline, here with $x_0$=½. See the convexity shape in Figure 7.6.

Now let us now continue with our example to solve for the variance of the **Y**, which is here really the variance of our financial account balance in one year. This approach is shown in Formula Z. So given the mathematics of solving for the variance with the series approximation, how would a visualization approach been useful? In order to see this, we look at the illustration below, with the aim to think through these problems in one's head, without resorting to a computer or calculator.

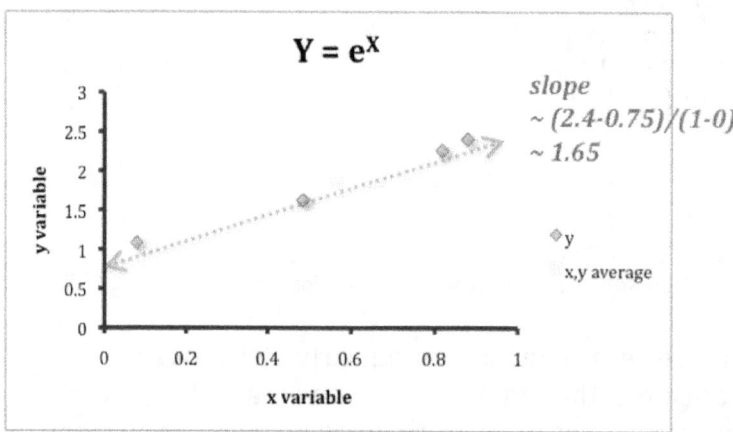

**Figure 7.8**

The variance of $Y$ is essentially the variance of $X$ times the square of the vertical $Y$ range it is mapped to. So we have 1/12*slope². This is 1/12*$e$, or ~0.2. We go through these visual steps in Formula AA, though the main point here is that we can solve this idea through multidimensional visuals. And without having to get bogged down in the power series math that got us to Formula Z.

## Chapter 8: Regressions

In this chapter we'll explore regressions in greater detail. There are three types of major regressions: linear, multiple, and multivariate. We discussed simple linear regressions earlier in this book. Cross-sectional analysis of economic data would fall into this category of linear regressions, and they generally does not have a time series element to them.

Multiple regressions are different than the simple linear regressions, only in that it deals with the analysis of multiple linear relationships between one dependent variable, and multiple other variables. Lastly the multivariate regressions are an analysis of a single relationship between one dependent variable and multiple independent variables paired with it. Note that the multivariate regression can not normally be visually charted.

It is also worth nothing that unlike with a linear regression, the choice of the dependent and predictive variable with multivariate regressions is more crucial versus with the simple linear regression. Even with linear regression, until one knows what the linear relationship is, it is best to chart twice by swapping the axis. This helps isolate dependent relationships in the data that might be easier to see under one variable designation versus the other (assuming the dependent variable is on the vertical axis, etc.).

A popular variation of these regressions above is when one of the independent variables is time. This is known as the science of econometric analysis. The linear regression analog in this science is named time series analysis. And a higher-level version of this is the multivariate regression analog, where the past dependent variables can also be used. When this is the case, we have what is called a vector auto-regression model. There are also cases where non-linear model using higher power terms (e.g., $X^2$) or interactive terms (e.g., $X_1{}^*X_2$)

Some of the basics of regressions that we've discussed earlier in the book are that one's dependent and independent variables would ideally be bi-normally distributed, with a large sample count of at least 30. But we should aim for possibly much more, as described later in the chapter.

Relationships between variables can generally be thought of as bivariate normal distributions. And it can be shown that this also implies that along any independent variable value, the dependent variable is identical and normally distributed about the regression line. This last point is useful for linear regression analysis, where we seek to minimize the linear deviations[2], as we noted earlier in the book, this was the way the science was built-up historically for solutions to use linear algebra. This was before the invention of computers that could now more easily do absolute, and orthogonal deviations.

Even with these basics satisfied, we should recall that there are two sources of errors in estimating a regression line: the intercept and the slope (particularly for the small amount of extreme independent data values). In Figure 8.1 below, we look at the relationship between Bank of America's price performance and the performance of the financial service sector, for about 30 days during 2011.

While leading departments crunch out these types of charts frequently, a more critical eye would generally prefer more data in some spots, and would be concerned that Bank of America represents such a large fraction of the financial sector (independent variable). [lii] Consequently the regression may not provide as much insight as the same chart for a smaller bank (*y*-axis) or the MSCI-World index (*x*-axis) would provide.

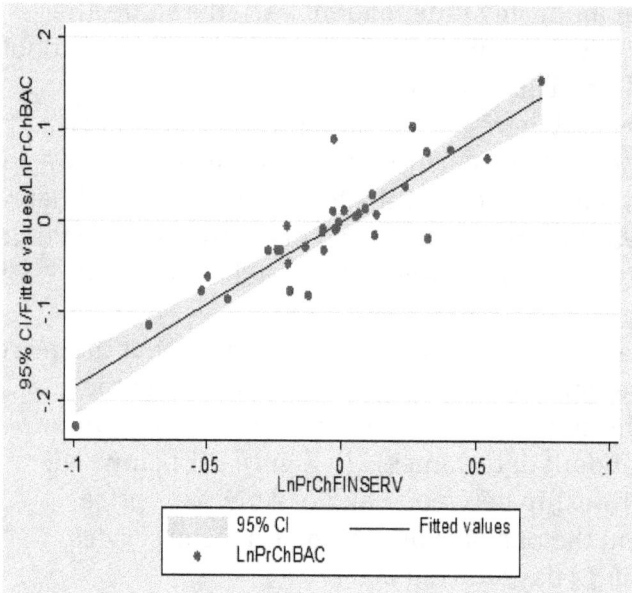

**Figure 8.1**

In this next example, we can see in Figure 8.2 multiple time series for a number of different U.S. national housing statistics that was carefully put into one chart using both vertical axis. And this is analogous the type of data sets that policy analysts and leading developmental bank economists look at given their nearly universal appeal. The time-series data highlighted in this 2009-2010 chart during the financial crisis clean-up are: monthly home asset values, new home sales, new home construction, foreclosure inventory, and existing home inventory.

As noted earlier, we can not display multivariate charts with a large number of variables. And it would even be difficult to have multiple variables' time-series in one chart, if those variables have very different scaled units.

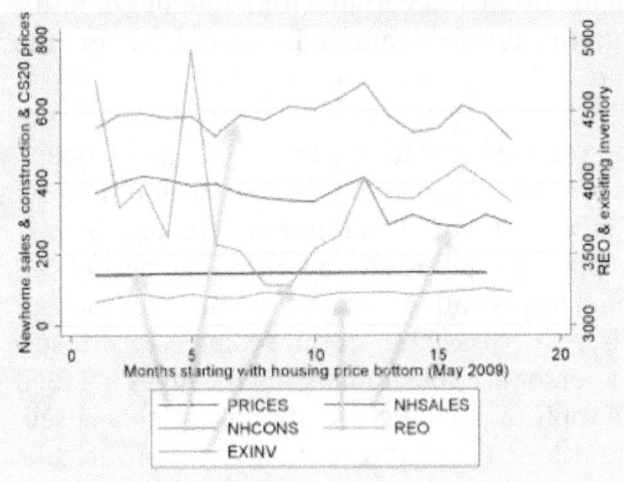

**Figure 8.2**

As noted before, colinearity can be an issue with a larger number of explanatory variables. It is difficult to consider multi-colinearity in multiple dimensions, though we have an example that gives us a view of this within a multi-media video.

http://statisticalideas.blogspot.com/2013/06/correlated-trivariate-normal.html

We also showed earlier how combinatorics can help solve the number of unique pairs of variables to be checked. For example, the number out of say 3 variables is $_3C_2=3!/(1!*2!)=3$, or out of say 5 variables is $_5C_2=5!/(3!*2!)=10$.

If we wanted to investigate the relationships with home asset values, we may want to start considering some combination of the other four variables as independent variables. And additionally, looking at the lagged values of those same independent variables for their auto-regressive properties.

The explanation of home asset values would benefit from a small number of explanatory variables, which are not colinear with one another. The whole model is termed a "vector auto-regression", with the sustained impact on one variable having its effect seen on the predicted variable for a number of future time periods, until this future strength dissolves back into the longer-term trend.

The key to a good regression is to use one's best independent variables, which are also not correlated with one another. So a correlation that is neither close to +1, nor -1. We do this by analyzing municipal debt spreads for a single time period. And have this explained by a swath of local economic data such as: budget short falls, real gross state product change, foreclosure rates, changes in personal bankruptcies, and changes in failed bank closings. A topic of partial-correlation can also be used to measure the knock-on correlation enhancement a variable provides, after accounting for the other independent variables correlation to it and the dependent variable. The mathematics for it are beyond the scope of this book but in my lecture slides.[liii]

We can see the two satisfying variables in the illustrations below, in Figure 8.2. As is often the case with economic data, we use log transformation of municipal government bond spreads in the U.S., versus the log transformation of two selected independent variables. We again notice the two illustrations share the same dependent *y* variable, along an identical vertical axis. Note that variable names shown from statistical software usually appears as one coded "word", without spaces or special characters. The natural log of **Y** could be shown as "**LnY**".

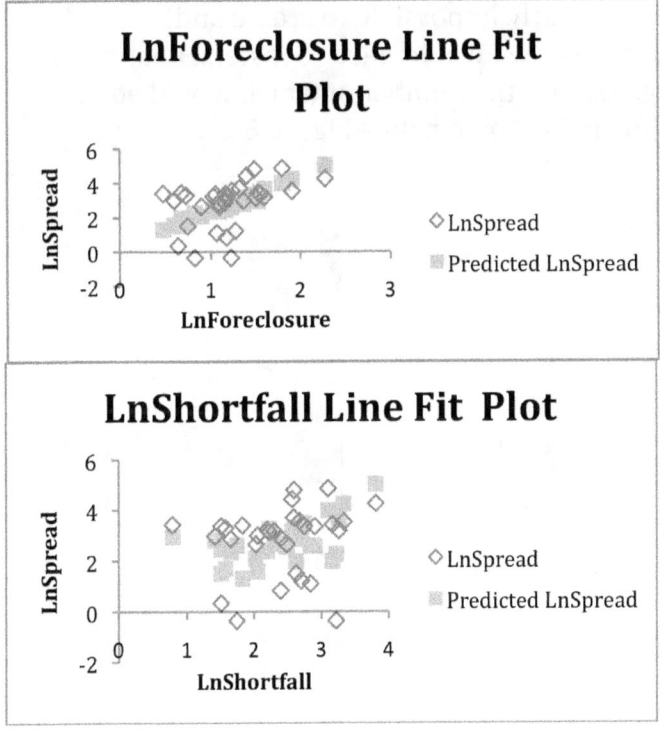

**Figure 8.3**

It can be seen that the state cross-sectional foreclosure data has more explanatory linkages to the municipal bond spreads than do the state-level budget shortfall. We can see this in the slope and the cluster about the regression lines. This allows the shortfall data to provide additional unique information, similar to how we suggested in the Merdeka Square example to ask witnesses for a nearby landmark, in addition to the location of the kidnapping.

So far we have shown an illustration about multiple variable time-series, and multiple variable regressions. In Figure 8.4 we show a demonstration of a multivariate of three dimensions (trivariate), which is now four dimensions and the practical limit of what any normal person can think about.

In this illustration below, we have identical normal distributions. We can see how a quadvariate normal distribution would be nearly impossible to create and understand. In such cases we would need to instead analyze multiple linear regressions, the number of which would equal the combinations formula shown below Figure 8.2.

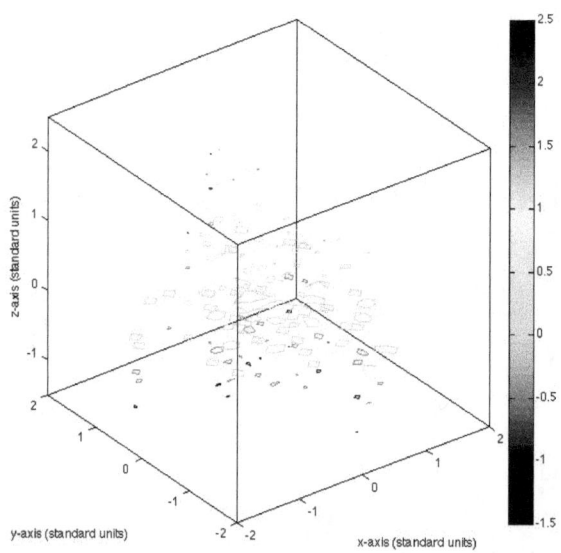

**Figure 8.4**

Note that a $\rho^2$ could also be computed here in the case of a multivariate regression, as we have done a couple of times earlier in this book, when we looked at simple linear regressions. But given the multiple independent variables, we need to assign a modification to the $\rho^2$ calculation. The modified $\rho^2$ takes into account the increased likelihood of being able to artificially overfit due to a larger number of variables relative to the sample size.

This adjusted-$\rho^2$ calculation is: $1-(1-\rho^2)*(n-1)/(n-k-1)$. Where the sample size is $n$, and the number of independent variables is $k$. As $k$ increases, the denominator of $(n-k-1)$ decreases, and hence the adjusted-$\rho^2$ also decreases.

Also for practical purposes when working with this sort of model, we would hope that the ratio of $n$:$k$ is greater than 4, with $n$ again best more than 30. Following these rules of thumb would help provide comfort in the quality of our multivariate regression match. Examples of their use are if $k$=2, the $n$ should be greater than 8. And if $k$ is 4, $n$ should be greater than 16.

A final example worth exploring concerns the topic of splicing. Splicing data allows us to acknowledge the changing probability distribution regimes as we move across the independent variable. See Figure 8.5 for actual versus hypothetical expected distribution for pension benefit data.

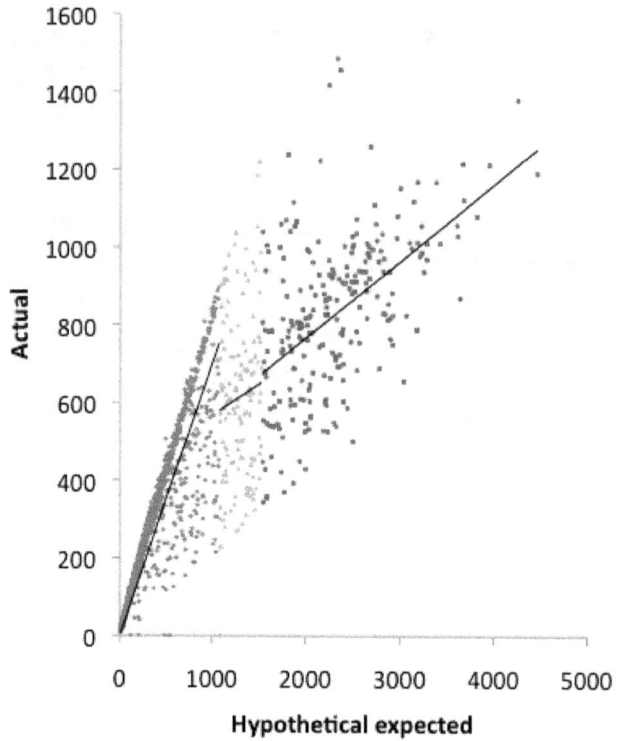

**Figure 8.5**

One can partition the data above or even invert it to better capture the economic trends. The point is that we are able to get stronger relationships at times, by splicing the relationship along the same independent variable.

Up to this point we have emphasized the need to take great care in analyzing the data. Even if computer and technology have met to allow organizations to rapidly crunch of deluge of noisy data (e.g., some Big Data initiatives), in many cases we've argued here in the book that it's our initial fundamental understanding of the small data and rational relationships that can make much difference.

## Chapter 9: Distribution fits

While we discussed parameter fitting earlier in this book, in this chapter we look at distribution fitting. With visual analysis leveraged as often as possible. Let's start by asking an essential question: what is it about our probability distribution that is important? Is it the middle, the ends, the thinnest points, the thickest points, the likelihood curvature at a specific percentile, etc.?

The tests differ for similarities or for contrasting distributions, and they depend on what sort of probability distribution we are working with. They also depend on the quantity and type of data (discrete or continuous distributions).

At the end of 2010, many citizens were interested in looking at the difference in size distributions for the bank failure that year and contrasting it with the same from 2009. Since the national bailouts, were things getting better for the banking industry then, and if so then by how much? Without getting to the bottom of these reporting questions from the data, we can't ask additional relevant questions, about the financial system then and its impact on the macroeconomy.

The Federal Depositary Insurance Corporation makes public the size data of failed banks, which is summarized in Figure 9.1 below.

|  | Average (natural log) | Variance |
|---|---|---|
| **2009** | $19.6 | 2.2 |
| **2010** | $19.3 | 1.7 |
| **Pooled years** | | 1.9 |

Figure 9.1

So down from $19.6, to $19.3. What a big improvement, right? Well perhaps not when we consider the statistical significance of these data, to also include its dispersion. We have two different variances, as shown, and so we consider a pooled variance ($s_p{}^2$). As shown in Formula AB, this variance is mostly the sample size weighted variances of each year, minus the degrees of freedom (mostly the number of banks per year) for the two years' sample size counts.

Of course this analysis would be easier is all the data were paired from 2009 to 2010, with no changes to the bank structures in-between. To make the statistics more difficult, we wouldn't be done with a pooled variance analysis against the means since we would be more cautious now on stray outliers tilting the results (while in paired analysis they might cancel). And this is somewhat more important for distribution fits here, as opposed to parameter fits we discussed early on in the book. The skew for the 2009 bank asset sizes is a large 1.8 (forcing the higher variance for that year), while the skew for 2010 bank asset sizes is near zero.

To better normalize for what's happening at the high end of the asset size range, we use Kolmogorov-Smirnov distribution fit technique. This technique diminishes the statistical impact of outliers. Here we show the percentile plot in Figure 9.2, for bank failures in both years.

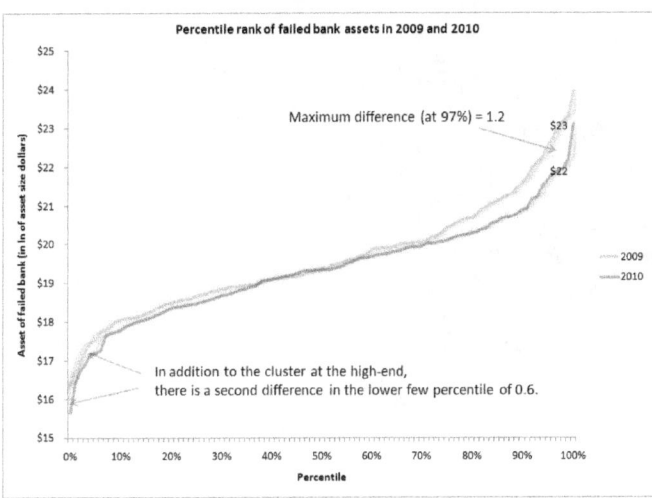

**Figure 9.2**

While we certainly see that the largest banks' failures in 2009 are larger than the largest banks' failures in 2010, we also see that these differences evaporate within the main part of the distribution, an area that better reflects the typical banks at risk then. So this must be part of the qualitative answer that the differences are not as great from a statistical perspective.

In running a two-sample Kolmogorov-Smirnov test with a critical value of 10% (i.e., 90% confidence), we mathematically show that there is no statistical difference between the asset size mix associated with the 2010 failures versus with the 2009 failures. This quantitative test examines the largest vertical difference between the cumulative percentile distribution plots.

Let's now explore another distribution fit test that is a new technique to many. In this next example, we explore whether three hypothetical sampled rest stops along the Silk Road are equally distributed. In Figure 9.3, we show these stops along this 6000-mile route, from ancient Peking, to Venice. One might guess ex-ante that these stops are not evenly distributed, but what would a distribution fit have to say about it?

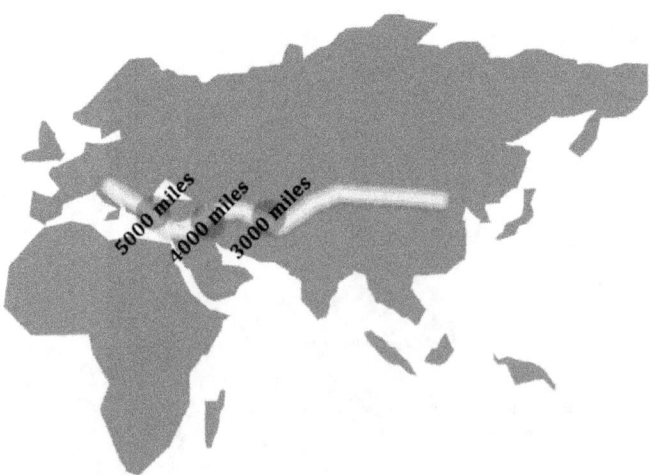

**Figure 9.3**

Now the three rest stops are as follows: **merchant$_1$**=3000 miles, **merchant$_2$**=4000 miles, and **merchant$_3$**=5000 miles. Given this data is discrete and generally not suitable for a cluster at one end or the other, the previous Kolmogorov-Smirnov test would not work. But we can try a new one that would be better, named the Anderson Darling statistic. The set-up in Formula AC is needed for this computationally intensive (only those with limitless free time would crunch this without a computer), but it is intuitively easy of an approach to grasp.

The low level of the $A^2$ statistic of 0.51, which measures the compounded differences in survival and probability functions between the two distributions, would not reject the idea that the sampled three stops are evenly spread across the 6000-mile journey. The Anderson Darling critical test at the 10% confidence level come to a variety range of somewhere between 0.6 and 1.9, depending on assumptions that we make about our confidence in the distribution model fit. In any event, the $A^2$ of 0.51 is less than the smallest end of that range, and we might assume that these rest stops were targeted by a uniform random variable.

In the last example we review the popular Chi-square ($\chi^2$) distribution fit technique. And we'll notice how it differs from the two other distribution fitting techniques we have so far discussed thusfar.

In November 2008, President Obama won a decisive U.S. election victory. And the Democratic political party gained seats in both houses of Congress. The 111$^{th}$ Congress came to session then, serving during the years 2009 to 2011. Something is less known about this time, but which we will solve for now. And that is which house of Congress had a larger Democratic party gain? See Figure 9.4, where we show the parties' count in each house during the 110$^{th}$ Congress, and then the 111$^{th}$ Congress. We can't simply look at just the absolute change in seats, or just the percentage change. Chi-square significance looks are a statistical combination of these two types of changes.

| Senate | 110th elected | 111th elected | | 111th expected | Chi-squared |
|---|---|---|---|---|---|
| Democrats (D) | 49 | 57 | | 49 | $(49 - 57)^2 / 49$ |
| Republicans (R) | 49 | 41 | | 49 | $(49 - 41)^2 / 49$ |
| *Total D&R* | *98* | *98* | | *98* | *2.6* |

| House | 110th elected | 111th elected | | 111th expected | Chi-squared |
|---|---|---|---|---|---|
| Democrats (D) | 233 | 256 | | 235 | $(235 - 256)^2 / 235$ |
| Republicans (R) | 198 | 178 | | 199 | $(199 - 178)^2 / 199$ |
| *Total D&R* | *431* | *434* | | *434* | *4.2* |

**Figure 9.4**

The slight difference in D&R counts from the 110th Congress, to the 111th Congress, do not matter for distribution fitting. This Chi-square analysis only uses D&R counts, and makes no assumptions about other members with other political affiliations, which were only five in total, between the two years. The degrees of freedom (**dof**) for both houses are the same, which is a reference to the count of both the number of Congressional years and number of political parties.

The $\chi^2$ data clearly show that the Senate had a much less significant increase in Democratic representation, in the 111th Congress. The distribution fit was more statistically significant with the House.

This may befuddle those who were focused only on the 16% pick-up in the Senate, versus the 8% pick-up in the House. Bear in mind that while this is a true difference, from a difference in percentage point of view, recall that the Chi-square formula also jointly considers that this is an easier hurdle from the near 50% expected base (look at the composition of the 110th Senate). It is actually much more difficult to achieve half the difference, as shown in the confidence interval figure earlier in the book, when closer to a diverse portion away from 50%.

For the Chi-square analysis, we generally insist on a minimal sample size of five in each table. The minimum in Figure 9.4 is in the 40s, so we are fine. Regardless, a larger lesson here is that isolated statistics output can be both correct, yet somewhat fickle. In this case we can focus on the fact that we are only comparing portions within one election year. To evidence this, we note that later with the 2010 mid-term elections, a powerful female politician Nancy Pelosi was forced to cede House leadership, showing how unstable the significant 2008 election result infers about the near term future of politics.

Just as there are different approaches to calculate variance (and sometimes to calculate a best fit line or even correlation), there are several techniques to get a $\chi^2$ statistic. The selection to choose is sometimes highly dependent the completeness and variation in the data or statistics we have. For example, there is one calculation variant, intended for data consisting of portions instead of numbers. Even in our case of having number of Congressional people, there are two variations as shown below.

| Chi-square, Figure 9.4 version | Chi-square, variation |
|---|---|
| *n = number of rows* | *n = number of rows* |
| | *o = number of observed values* |
| | |
| *data$_i$* | *data$_i$* |
| *= (observation$_i$ – expected$_i$)$^2$ / expected$_i$* | *= observation$_i^2$ / expected$_i$* |
| | |
| | *l = sum all data of i, from 0 to n* |
| | |
| *$\chi^2$ = sum all data of i, from 0 to n* | *$\chi^2$ = l - o* |

When looking up significance of a $\chi^2$ statistic, a **dof** measure also needs to be calculated: (**n** –number of estimated parameters -1)*(columns -1). If either the number of columns or **n** (rows) is one, then we ignore the what's in the (...) for the left or right side of the "*" in the earlier sentence. If we have one row, then the number of columns minus one is the **dof**. If both rows and columns equal one, then there is nothing to solve as clearly no $\chi^2$ is possible. One is encouraged to try a computer simulation of their own.[liv]

Also we see through a mid 20th century theorem, named after one of Harvard's first statisticians, William Gemmell Cochran, that the sum of independent normal distributions (e.g., one per number of categorical rows) is equal to a $\chi^2$ distribution.

What's more in regards to $\chi^2$, is that a sum of independent $\chi^2$ distributions also retains the $\chi^2$ distribution characteristics (not a normal distribution some might just throw out). It is assumed of course that these independent $\chi^2$ distributions are aligned with one another.

So now here is our final distribution mapping for the models we have discussed in this book:

*l.d.         = limiting distribution*
**i.d.         = independent distributions**

*Bernoulli* $\rightarrow_{l.d.}$ *binomial* $\rightarrow_{l.d.}$ *normal*
*normal* $\leftarrow_{l.d.\ with\ expected\ fit\ \rightarrow\infty}$ *Poisson* $\leftarrow_{l.d.\ with\ sum\ of\ i.d.}$ *Poisson*

***Chi-square*** $\leftarrow_{l.d.\ with\ sum\ of\ i.d.}$ ***Chi-square*** $\leftarrow_{l.d.\ with\ sum\ of\ i.d.}$ *normal*

In addition, we show a mapping below for a number of other probability distributions that were not covered in detail in this book. The parameters for these distributions can be identified by the parentheses shown after the distribution name. We also see that it is only for the negative binomial, that we need to employ the limiting concept in order to morph from one distribution, into another. [lv] Limiting approaches were more common in the many probability distributions we discussed here in the book, and shown in the mappings just before this paragraph.

*model of variable named X          = X~model*
*model of variable named Y          = Y~model*
*model of variable named Z          = Z~model*
*sum of i.d. models                 = S~model*

*X~Weibull($\theta,\tau$)* $\rightarrow_{\tau=1}$ *X~exponential($\theta$)* $\leftarrow_{\alpha=1}$ *X~gamma($\alpha,\theta$)*
*X~gamma($\alpha,\theta$* $\rightarrow_{\theta'=1/\theta}$ *1/X~inverse gamma ($\alpha,\theta'$)* $\rightarrow_{\alpha=1}$ *1/X~inverse exp.($\theta'$)*

*S~neg. binomial ($\beta,r$)* $\leftarrow_{l.d.\ with\ r\rightarrow\infty}$ *Y~neg. binomial ($\beta,r$)* $\rightarrow_{r=1}$ *Y~geometric ($\beta$)*

*beta(a,b,l.b.,u.b.)* $\rightarrow_{with\ a=1,\ b=1}$ *uniform (l.b.,u.b.)*

*Where l.b. is lower bound, and u.b. is upper bound.*

The ultimate message of this chapter is that in a practical setting we would have a rich intuition for the types of probability models that can be used, and combine this with some established rules of thumbs to provide us with a general guide.[lvi] We discussed in recent chapters that we should attempt to try different parsimonious models to know the strengths and weaknesses of different ones for a given problem, and spend more time thinking through the logic of statistical relationships, as opposed to model fitting.

For the distribution fitting statistical tests covered in this chapter, all three follow the test criteria that a higher test statistic increases the chance that our distribution fit will not be accepted. Now we'll move from distribution fitting techniques, towards looking at popular probability distributions.

What probability models have the most appeal, as measured by Google or "bing" internet searches? We might think that the most applied models throughout the world would have the most search results. And most may guess correctly that this probability model is for the normal distribution.

But which probability models would come up next, after the normal? See the Google results in Figure 9.5. In addition to the count, we highlight which probability models also have the most images or pictures associated with them through the internet.

| Distribution | Search hits count (in millions) | Most pictured |
|---|---|---|
| normal | 25.4 | Yes |
| uniform | 10.1 | |
| beta | 7.0 | Yes |
| exponential | 3.6 | |
| inverse gamma | 3.0 | |
| gamma | 3.0 | |
| Poisson | 2.8 | Yes |
| binomial | 1.6 | |
| negative binomial | 0.7 | |
| Weibull | 0.6 | |
| Gompertz | 0.1 | |

**Figure 9.5**

This may be a surprise on one level, for those reading this book, to see a seeming over-reliance in the use of the normal distribution to explain phenomena. This slight counter-intuition about not appreciating higher-order moments, also are evidenced in the subsequent few distributions in the ranked list above.

What these popular distributions (normal, and uniform) have in common are that they are mapped with a family of models, and are also both symmetrical, and continuous. What's also perhaps interesting is how far down the Poisson is. While it is frequently viewed, some may not appreciate that these frequency models are a result of a Poisson process that looks at a continuous exponential time duration between events (exponential being much higher on the list versus Poisson).[lvii]

We can also try to think about what a search-weighted mixture of the distributions might look like, and imagine a smoothed down super distribution result that is also somewhat unique. Recall from early on this book that if distributions were like beverages, then we are not looking for an Arnold Palmer. An example insted, which is also not simply a mixture of distributions of the earlier page, is similar to the shape of the top of a fractal snowflake. Except we just follow the marginal distribution weights of this shape, which has a large mounded middle and smaller multimodal peaks flanking it.

In Figure 9.6, we show this snowflake fractal, most commonly attributed to the early 20th century Swedish mathematician Niels Fabian Helge von Koch. We again focus chiefly on the upper portion of this symmetrical fractal, as we show above the dashed line. Fractals are interesting choices for shapes as they have patterns that often replicate in nature, such as the curvature of everything from shorelines, to sea shells.

http://statisticalideas.blogspot.com/2013/09/fractals-and-probability.html

They also show up in probability models, such as those that define risk boundaries, as shown in the above linked simulation of a fractal probability.

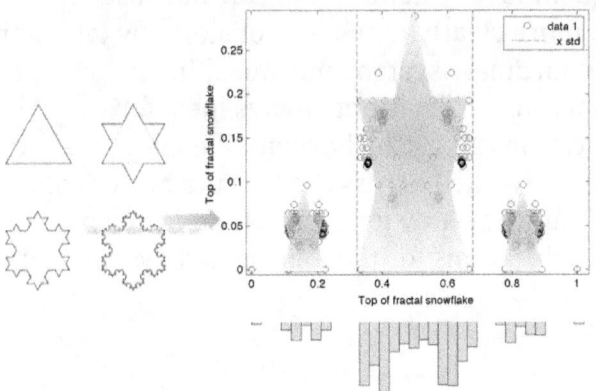

**Figure 9.6**

On the right side of the representation, we generate thousands of computer simulations of this portion of the fractal snowflake. Fractals are boundaryless by definition, so the illustration is merely an approximation of the resulting distribution. The key is that we can then collapse the weight of the vertical marginal density (thin vertical probability slices along the **x** axis), and see this distribution (inverted) on the lower right illustration.

The discontinuous nature of the fractal snowflake on top ensures the uniqueness of the resulting probability distribution on the lower part of the chart. And through this procedure we get the curved probability pattern desired: a normal distribution type of central mound, flanked by small modal weights that are usefully clustered at the inner quartiles.

As a final topic we discuss visual ways to show a probability distribution. Similar to the difficulty of showing multivariate regressions, probability distributions are not bounded by just **x** and **y** axis. For example, in Figure 9.7, we show the distribution of growth across the globe. The 17th century metaphysicist, René Descartes, whose essays on the subject have been recently translated, established what we see as the Latin Cartesius geometric system.[lviii] For our purposes we can now see the data more clearly in two dimensions, by taking the global Cartesian coordinate system and smashing it down into a two dimensional chart. This is not always ideal for understanding locations along the horizontal edges, where the polar cap has been stretched across the top and bottom of the chart. Note this is anyway how many media reported maps of possible flight path locations for the missing Malaysian Airlines flight in early 2014.

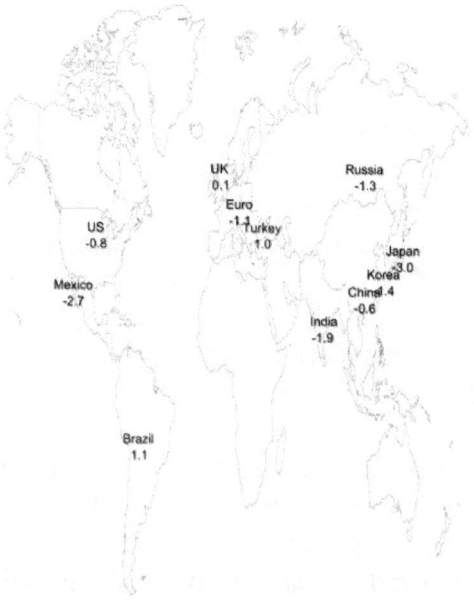

**Figure 9.7**

There are multiple distributions that can be shown here. We could explore the difference in global growth over time. Or show the chart with a moving time dimension, which would rely less on numerical labels in order to visually understand it. Over time and with practice, we'll see more creative ways to think about distributions in a visual way.

http://statisticalideas.blogspot.com/2013/07/long-term-growth-and-unemployment.html

## Chapter 10: Other topics, and summary

I hope that you have enjoyed this book and gained important insights into both the power, and the fragility of statistical models. Many of the interesting topics that we come across in life involve exploring higher-order dimensions to a data set or distribution. It is wonderful to be able to understand and communicate experiences with through this prism as well. For example, if we were an insurance company and wanted to underwrite a storm damage policy for the next three days, would the average weather be all we need to know, or would the dispersion within the three days and other characteristics also be helpful? Clearly we want to have some amount reserved for a risky occurrence at some point, so understanding the dispersion and when it might be triggered would be important.

Also note some basic rules, such as for summarizing information. It is good to try to collect a large sample size in most cases, and also to be able to pick a couple different summary statistics in order to describe a probability distribution or even a parameter (e.g., median and mode for a central tendency). We should have a healthy sense of uncertainty when analyzing information, as we have seen differences in calculations exist even on the same data, for example when parameter matching or thinking about things like correlation or sampling assumptions. So be able to discuss the areas where your conclusions are strong and what the weaknesses are. And be able to describe what adjustments may need to be made to data or analysis to better reflect the main thrust of reality.

For fun, let's look at Figure 10.1, and guess which forecaster is a layperson and which is a probabilist:

| Forecaster | Today | Tomorrow | Day after tomorrow |
|---|---|---|---|
| A | 100% chance of rain | A hurricane striking town is guaranteed | Pretty sure that it'll be sunny all day |
| B | 50% chance of rain | 60% chance of thunderstorms | 70% chance of cloudy skies, unless day prior it is dry |

**Figure 10.1**

Forecaster B is clearly more of a measured probabilist. Sometimes painfully precise in their mathematical language, yet it is for good reason.

In addition to summarizing data, we should consider the impact of transformations, such as logs or portions, or applications of data splicing. Even for a basic statistics such as the average, the consideration of geometric versus arithmetic is interesting. Take an annuity balance over two periods of {1,2}. The arithmetic average is 1.5. But the geometric average, which incorporates convexity, is less at $\sqrt{(1*2)}=1.4$. In investment finance, we come across this issue in places such as multi-temporal models, where one can use the arithmetic returns in the Sharpe risk-adjusted ratio, and the geometric returns for the Capital Asset Pricing Model.

We also see in this book that probability distribution functions have unique fingerprints that differentiate one from another. Look at the lissome, probability curve in Figure 10.2 below. Is this from a lognormal, binomial, geometric, Poisson, or other distribution?

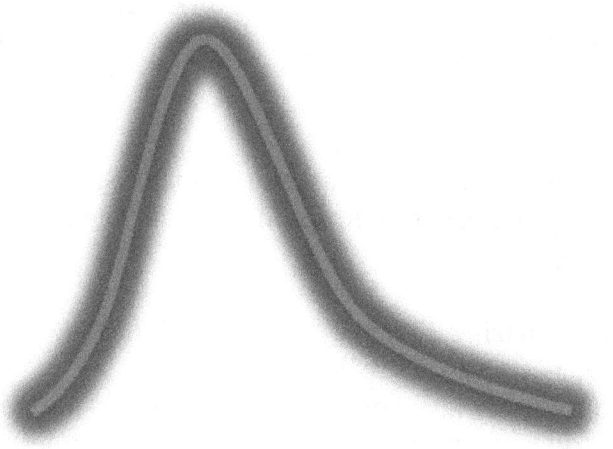

**Figure 10.2**

The moment generating functions can help unlock this unique "fingerprint". And even with it, we are not excused from downstream analysis of the distribution shape that requires additional parameter analysis. During the financial crisis, and more recently with J.P. Morgan's "London Whale" debacle, we see the importance of investors understanding the tail risk parameters of seemingly straight models, such as value-at-risk (VaR).[lix]  This shares the acronym but is different from "vector auto-regression" we discussed earlier in the book.

VaR is an interesting risk measure given CEO Dimon preferring to not use this very popular risk measure invented at his firm decades earlier. For its benefits, VaR does suffer from some lack of coherent properties of probability theory. So this makes looking at multiple, more complex, risk statistics measures all the more vital.

The coherent risks satisfy all these four properties, and all of these we do not need to understand in great detail here: sub-additivity, translation, positive homogeneity, and monotonicity. Sub-additivity is the property that doesn't fit since it can be gamed with two risks that are valued each at just below the VaR radar in isolation, but together are a critical risk.

An interesting next query is if we used other distributions for risk modeling, such as the generalized extreme value (GEV) distributions, then we'd be able to explore a deeper set of fat tailed risks that better capture tail risk events. Higher order risks, which don't show up as nicely through fractal geometry, though they do a good job of pointing to the sorts of Pareto distributions that are better than the normal distribution in looking at risk models.[lx]

As a final note, I hope that you are able to better appreciate some exciting, new statistics ideas and are able to express them better in your life. With continuous perusing back at this book, you'll appreciate the complexity in it will keep unlocking ideas and tool, in addition to the logical underpinnings supporting them. Also we showed that visual techniques could be used to think through probability models, which we have argued many times as much of an art that needs to be understood, as they are science. Please continue to work with probability and statistics, and participate by following the current literature on the same. Ask me any questions you come across.

Thanks much for sharing this book experience (a dozen years in the making) with me.

# Formulas
## Formula A
$\sigma_{sum}^2$ = average event count * (average$_{severity}^2$ + $\sigma_{severity}^2$)

*Unlike with the average sum of count and severity, which just looked at average severity, here one sees in the above function that we look at more than just $\sigma_{severity}^2$.*

*Now we can also verify that this above formula collapses to our traditional Poisson process:*
$\sigma_{sum}^2$ = average event count * $(1^2 + 0^2)$
= average event count, or variance of the Poisson equals the average

## Formula B
*Finding 1 coin in the first 2-hour period:*
$e^{-\lambda}*(\lambda^k)/k!$        = $e^{-\lambda}*(\lambda^1)/1!$
            = $e^{-\lambda}*(\lambda)$

*Finding 3 coins in the next hour:*
$[e^{-\lambda}*(\lambda^{2k})/(2k!)]^{\frac{1}{2}}$     = $[e^{-\lambda}*(\lambda^6)/6!]^{\frac{1}{2}}$
            = $[e^{-\lambda}*(\lambda^6)/6!]^{\frac{1}{2}}$

*Note that an hour is ½ of a 2-hour period, so we use "2k" instead of "k" in the second formula, as modeling the "$k^{th}$" coin for 1 hour is the same rate as 2k-th coins for a 2-hour period.*

## Formula B2
n                 = sample size

variance$_\theta$         = $n * \theta^2 / (n + 1)^2 / (n + 2)$
                 = $3 * 4^2 / (4^2 * 5)$
                 = 3/5, or 0.6

## Formula C
*Say we wanted to know the probability of event A, conditioned on event B:*

$p(A|B)*p(B)$     = $p(B|A)*p(A)$
$p(A|B)$         = $p(B|A)*p(A) / p(B)$

*While the general formulation is shown above, sometimes we need to further re-describe p(B) to attain a mathematical solution.*
$p(B)$            = $p(B|A)*p(A) + p(B|A \text{ complement})*p(A \text{ complement})$

## Formula C2

*expected value of holding out at 2pm*
*= expected value of 3pm movie*
*= average (3 stars, 1 star)*
*= 2 stars*

*expected value of holding out at 1pm*
*= expected value of 2pm movie decision*
*= average [5 stars, max. (1 star, or the expected value of holding out at 2pm)]*
*= average [5 stars, max. (1 star, 2 stars solved for above)]*
*= average (5 stars, 2 stars)*
*= 3.5 stars of holding out at 1pm > 3 stars of watching the 1pm movie*

## Formula D

*Since the borough population can only be >1m, or <1m, we can convert the population dependent variable to an indicator value of "1" for populations >1m, or "0" otherwise.  Let's name a "super variable" X.*

$X = intercept + b_1 * wages + b_2 * area + residual error$

*X is now a regression that can theoretically be anything from -∞, to (+)∞.*

*e is the base of the natural logarithmic numbering system, with a value of ~2.7*

$e^X$ *is therefore any value between 0 and ∞.*

$e^X / [1 + e^X]$ *is therefore mapped to any value between 0 and 1.*

## Formula E

$$p \qquad = e^X / (1 + e^X)$$

*Which after some algebraic gymnastics, one can see would equal the following.*

$$\ln[p / (1 - p)] \qquad = X$$

*So we know that our logistic regression X would map to probability we are hoping to solve for:*

$$\ln[p / (1 - p)] \qquad = intercept + b_1 * wages + b_2 * area + residual\ error$$

*This is the logistic regression model that solves for the model coefficients (e.g., $b_1$, $b_2$, etc.) and provides an estimate for the missing value that is between the logical boundaries between 0 and 1.*

| | |
|---|---|
| intercept | = -181 |
| wages | = 0.1 |
| area | = 1.6 |

## Formula F

$$\ln[p / (1 - p)] \qquad = -181 + 0.1 * 800 + 1.6 * 60 + 0$$
$$= -5$$

$$p\ estimate \qquad = e^X / (1 + e^X)$$
$$= 10\%$$

*Note this 10% p estimate of is closer to the 0 indicator for borough populations < 1m.*

## Formula G

| | |
|---|---|
| (standard deviation) | < ½ * variance |
| 2 * (standard deviation) | < (standard deviation)$^2$ |
| 2 | < standard deviation |
| 4 | < variance |

## Formula H

variance$_X$ $\qquad = average[(X's - average_X)^2]$

Which after some algebraic manipulations we see:

variance$_X$ $\qquad = average(X's^2) - (average_X)^2$

So now we continue with the variance, and the coefficient of variation equations:

$(CV_X)^2$ $\qquad = [average(X's^2) - (average_X)^2] / (average_X)^2$

$\qquad\qquad = average(X's^2) / (average_X)^2 - 1$

## Formula I

Variance of n times a distribution $= average[(n * X's)^2] - (n * average_X)^2$

$\qquad\qquad\qquad\qquad\qquad = n^2 * average(X's)^2 - n^2 * (average_X)^2$

$\qquad\qquad\qquad\qquad\qquad = n^2 * variance_X$

If X has a uniform probability distribution, bounded by (min of a, max of b)

$CV_X$ $\qquad = 1 / \sqrt{3} * (b - a) / (a + b)$

## Formula J

average$(X^2)$ $\qquad = sum\ over\ all\ i\ [i^2\ e^{-\lambda}\ \lambda^i / i!]$

$\qquad\qquad = \lambda * sum\ over\ all\ i \geq to\ 1\ [i\ e^{-\lambda}\ \lambda^{i-1} / (i-1)!]$

So we have shifted values to the right one unit, because the new "i" is referencing the original Poisson distribution probability weights that were associated with "i-1".

## Formula K

XR $\qquad = ranking\ of\ the\ X\ variable\ values\ \{x_1, x_2 ... x_n\}$

YR $\qquad = ranking\ of\ the\ Y\ variable\ values\ \{y_1, y_2 ... y_n\}$

As an example, say we have a sample size of four. And if the X variable value set of $x_1, x_2, x_3, x_4$ is {5, 6, -99, 0}, then the XR variable value for X is $xr_1, xr_2, xr_3, xr_4$ of {3, 4, 1, 2}.

These XR's and YR's are then used for the Spearman's rho:

N $\qquad = sum\ over\ all\ n\ samples\ of\ t\ (xr_t - yr_t)^2$

$_sr_{XR,YR}$ $\qquad = 1 - 6*N/[n*(n^2-1)]$

Note that if the underlying distribution is a uniform, then the Spearman and Pearson are equal.

## Formula K2

*We then look at all pairs of $X_i$, $Y_i$ and $X_j$, $Y_j$ where $i \neq j$ and both are between 1, and n.*

$p_c$       *= all pairs where $X_i - X_j$ and $Y_i - Y_j$ are of same sign (i.e., concordance)*
$p_d$       *= all pairs where $X_i - X_j$ and $Y_i - Y_j$ are of different sign*

*Then we have Kendall's tau:*
$t_{X,Y}$     *= $(p_c - p_d)/[(n(n-1)/2]$*

*The Spearman and Kendall nonparametric rank correlations are of course connected, but not in an intuitive and linear way. The weighting in the equation above shows that Kendall $\tau$ tends to be relatively closer to zero:*

$3/2*\tau - \frac{1}{2}$       $<_s\rho$       $\leq \frac{1}{2} + \tau - \frac{1}{2}\tau^2$,    *when $\tau \geq 0$*
$-1/2 + \tau + \frac{1}{2}\tau^2$    $\geq_s\rho$       $> 3/2\tau + \frac{1}{2}$,       *when $\tau < 0$*

## Formula L

$p$                 *= probability of going right (r)*
$q$                 *= probability of going left, which we'll set to $\frac{1}{2}$*
$m$                *= row number*

*First solving for Figure 4.2:*
$variance_m$      *= $\Sigma$ over all row positions $[_mC_r * p^r * q^{m-r} * (r- (m - r))^2]$*
                      *= sum over all row positions $[_mC_r * (2 * (r - m/2))^2] / 2^m$*
                      *$\rightarrow$ m (by deductive observation of Figure 4.2)*

*Then solving for related binomial:*
$variance_{binomial}$    *= sum over all row positions $[_mC_r * (\frac{1}{2} * 2 * (r - m/2))^2] / 2^m$*
                      *= $m * p * q$*

## Formula L2

*For z, say of 0.05:*
*Move to an empty cell.*
*Enter "=NORMSDIST(0.05)".*
*Press return.*
*See "0.52" in same cell.*

## Formula L3

*What variance$_{binomial}$ would align to the variance$_{Monte\ Carlo}$?*

$$variance_{binomial} = [range]^2 * p * q$$
$$= [1 - (-1)]^2 * ½ * (1 - ½)$$
$$= 2^2 * (½)^2$$
$$= 1$$

## Formula M

*Say we have one variable X, which has three possible values $\{X_1, X_2, X_3\}$, which equals $\{4, 8, 2\}$. For aggregating statistics, does order matter, when we sample two of the three?*

<u>For a combination, order doesn't matter:</u>
$$_3C_2 = [3!] / [(2)! * (3-2)!]$$
$$= 3\ combinations$$

| average$\{X_1, X_3\}$ = 3, | average$\{X_2, X_3\}$ = 5, | average$\{X_1, X_2\}$ = 6 |
|---|---|---|
| grand average = 14/3, or 4.7 | | |

<u>But in a permutation, order does matter:</u>
$$_3P_2 = [3!] / [(3-2)!]$$
$$= 6\ permutations$$

| average$\{X_1, X_3\}$ = 3, | average $\{X_2, X_3\}$ = 5, | average $\{X_1, X_2\}$ = 6 |
|---|---|---|
| average $\{X_3, X_1\}$ = 3, | average $\{X_3, X_2\}$ = 5, | average $\{X_2, X_1\}$ = 6 |
| grand average = 4.7 | | |

But in any case, the ordering does not matter for the aggregating statistics distribution.

## Formula N

*From the language of Brownian motion....:*

$z$        = dispersion as a function of time

$S$        = distance from storm center

         = drift * time + $SD_{storm\ dispersion}$ * $\sqrt{(time)}$

*...to the language of stochastics:*

$dS_z$      = change in S per unit change in z, similar to elasticity
         ~ delta

$dS_t$      = change in S per unit change in time, similar to drift
         ~ theta

$dS_{zz}$    = acceleration, or change in the "change of S per unit change in z"
         ~ gamma

$dS_{tt}$    = acceleration, or change in the "change of S per unit change in time"
         = 0

We get change in S or "delta S" by taking a derivative with respect to time and a derivative with respect to z, and collect those chained time and z terms:

delta S   $\approx (dS_t + dS_z) * \Delta$ in time   + $(dS_t + dS_z) * \Delta$ in z
         $\approx (dS_t +$ "something"$) * \Delta$ in time   + $(0 + dS_z) * \Delta$ in z

## Formula O

$dD_z$                = change in damage per unit change in z
                   = -5

$dD_S$                = change in damage per unit change in S
                   = 20 * S

$dD_{zz}$             = change in the "change of damage per unit change in z"
                   = 0

$dD_{SS}$             = change in the "change of damage per unit change in S"
                   = 20

## Formula P

TFC = *true FC*

SFC = *sample FC*

*TFC average* $\pm y_p * (SD_{SFC\ average})$  $\leq$ *TFC average + (5% of TFC average)*

*Ignoring the TFC average, one wants a certain level ($y_p$) of one's sample FC variation to stay within 5% of the average TFC. This certain level ($y_p$) will be akin to the critical t-value. Continuing with inequality rearrangement:*

| | |
|---|---|
| $\pm y_p * (SD_{SFC\ average})$ | $\leq$ *(5% of TFC average)* |
| $\pm \sqrt{(variance_{SFC\ average})}$ | $\leq$ *(5% of TFC average)/$y_p$* |
| $variance_{SFC\ average}$ | $\leq$ *[(5% of TFC average)/$y_p$]$^2$* |
| $variance_{SFC}$ /*sample size* | $\leq$ *[(5% of TFC average)/$y_p$]$^2$* |
| $variance_{SFC}$ /*(TFC average)$^2$* | $\leq$ *[5% / $y_p$]$^2$ * sample size* |

## Formula Q

$variance_{number}$
$$= p * q * n$$
$$= p * (1 - p) * n$$

**$variance_{cumulative\ probability}$**
$$\approx [p / n] * [(1 - p) / n] * n$$
$$\approx [n * p / n] * [n * (1 - p) / n] * n / n^2$$
$$\approx n_x * (n - n_x) / n^3$$

One will also notice that the first equation above could be converted to *"p\*(1-p)/n"* fairly easily, which is then *variance(number)* divided by $n^2$. If one doesn't see this, then here it is again below.

$variance_{cumulative\ probability}$
$$\approx [p / n] * [(1 - p) / n] * n$$
$$\approx [p] * [(1-p)] * n / n^2$$
$$\approx variance\ (number) / n^2$$

# Formula R

$U$

$$= exp^{[z \, * \, \sqrt{(Variance(1-p) \, / \, (1-p) \, / \, \ln(1-p))]}}$$
$$\approx exp^{[1 \, * \, \sqrt{(0.0134 \, / \, (35\%) \, / \, \ln(35\%))]}}$$
$$\approx exp^{[-.314]}$$
$$\approx .73$$

*We solved above for a one standard deviation change from this delta method. Then inverting the survival function here we can reflexively obtain the cumulative probability function we sought.*

$\{(1-p)^{1/U}, (1-p)^{U}\}$

$$= \{(p)^{1/U}, (p)^{U}\}$$
$$\approx \{35\%^{1/0.73}, 35\%^{0.73}\}$$
$$\approx \{24\%, 46\%\}$$

*See how this contrasts with the earlier symmetrical CI of $35\% \pm 12\% = \{23\%, 47\%\}$*

# Formula S

*penalty*

$$= n \, / \, (n - 1)$$
$$= 7 \, / \, 6$$

*z-statistic for 90% CI*

$$= 1.645$$

*This z-statistic could be retrieved from the abridged table shown earlier in the book, though soon one should be know this constant value as second hand information.*

*t-statistic*

$$\approx \text{z-statistic} \, * \, \text{penalty}$$
$$\approx 1.64 \, * \, (7/6)$$
$$\approx 1.9$$

$CI_{distance}$

$$= \pm \, \text{t-statistic}_{distance} \, * \, \sqrt{(variance)}$$
$$= \pm \, \text{t-statistic}_{distance} \, * \, (standard \; deviation_{distance})$$
$$= \pm \, 1.8 * (.08)$$
$$= \pm \, 0.15 \; miles$$

## Formula T

*Marginal density of x in the bi-normal distribution of two independent variables:*

$$f_X(x) = 2 / \pi \sqrt{(1 + x^2)}$$

*and similarly, marginal density of y in the bi-normal distribution of two independent variables:*

$$f_Y(y) = 2 / \pi \sqrt{(1 + y^2)}$$

*One can then scale this for any size of circle. For a radius of 2, instead of 1, this implies a horizontal stretching of the circle by a factor of 2. So one's new $f_X(x)$ is half (or one divided by the factor) of the one solved for here.*

## Formula U

$w(out)$:
$\approx \pi * average[\phi(0), \phi(0.11/0.15*1.9), \phi(0.11/0.15*1.9)] * [average(0.11, 0.15, 0.15)]^2$
$\approx average[4.4, 1.6, 1.6] * (0.06)$
$\approx 0.14$

$w(in_{longitude \ or \ latitude})$
$\approx (0.11*0.04) * average[\phi(0.11/0.15*1.9), \phi(0.11/0.15*1.9), \phi(1.9)]$
$\approx (0.05) * average[1.1, 1.1, 2.1]$
$\approx 0.01$

$w_{total}$
$= w(in)_{longitude} + w(in)_{latitude} + w(out)$
$\approx 0.14 + 0.01 + 0.11$
$\approx 0.16$

*Where $\phi$ is the normal continuous density function.*

$p(out)$            $= w(out) / (w_{total})$
                               $\approx 92\%$

$p(in)_{longitude+latitude}$      $= [w(in)_{longitude} + w(in)_{latitude}] / (w_{total})$
                               $\approx 8\%$

$p(within \ CI)$           $= [p(out) + p(in)_{longitude \ or \ latitude}]^2$
                               $\approx [96\%]^2$ *(same logic as in text before Figure 6.3)*
                               $\approx 91\%$ *(or ~90% probability interval we solved for)*

## Formula U2

*cone volume + cylinder volume*

$= (1/3)\pi*(r)^2* ht_{cone} + \pi*(r)^2*ht_{cylinder}$

$= \pi * (r)^2 * [(ht_{cone\ starting\ at\ top\ of\ cylinder} - ht_{cone\ base\ at\ top\ of\ cylinder})/3 + cylinder\ ht]$

$\approx \pi * (r)^2 * [(4.4 - 1.1) / 3 + 1.1]$

$\approx \pi * (0.15)^2 * [(4.4 - 1.1) / 3 + 1.1]$

$\approx 0.16$, *or about* $w_{total}$

*Where* **ht** *represents height.*

## Formula V

<u>*Linear regression model for confidence:*</u>

*height* $\quad = level + slope * age + error$

$S_{xx}$ $\quad = sum\ over\ all\ individuals\ (person's\ age - avg.\ age)^2$

*SE of slope* $\quad \rightarrow$ *SE of height* $/ \sqrt{(S_{xx})}$

*+SE of level* $\quad \rightarrow$ *SE of height* $* \sqrt{[n^{-1} + avg.\ age^2/S_{xx}]}$

*=SE of height* $\quad \rightarrow$ *SE of height* $* \sqrt{[n^{-1} + (person's\ age - avg.\ age)^2/S_{xx}]}$

*Where* **n** *represents sample size.*

## Formula V2

<u>*Linear regression model for prediction:*</u>

*height* $\quad = level + slope * age(new) + error$

*SE of new* $\quad = SE\ of\ height * \sqrt{[1/ + 1/n + (age\ of\ new - avg.\ age)^2/S_{xx}]}$

$\quad\quad\quad \approx 0.3 * \sqrt{[1 + 1/9 + (11 - 6)^2/ 60]}$

$\quad\quad\quad \approx 0.4$, *versus 0.3 from the main age 2 through 10 data*

*We then divide by* $\sqrt{250}$ *if we want the SE of the new data's avg. height.*

## Formula W

*SSR/SST* $\quad = (\beta_1 SS_{xy})/SS_{yy}$

$\quad\quad\quad = [Cov(x,y)/(\sigma_x)^2 * Cov(x,y)*n]/[(\sigma_y)^2*n]$

$\quad\quad\quad = [Cov(x,y)^2*n] / [(\sigma_x)^2(\sigma_y)^2*n]$

$\quad\quad\quad = Cov(x,y)^2/(\sigma_x)^2(\sigma_y)^2$

$\quad\quad\quad = [Cov(x,y)/(\sigma_x*\sigma_y)]^2$

$\quad\quad\quad = \rho^2$

## Formula X

*The linear correlation is:* $\quad\quad\quad\quad\quad\quad$ $\rho = \sqrt{((5.25-3.7)/5.25)}$ $\quad$ =0.5
*While the orthogonal line correlation is:* $\quad$ $\rho = \sqrt{((5.25-5)/5.25)}$ $\quad\quad$ =0.2

## Formula Y

*x* $\quad\quad\quad\quad\quad\quad\quad\quad\quad\quad$ = *future age for a target individual*

*Makeham survival to age (x)* $\quad\quad$ $= e^{\wedge}[-A * x - m * (c^x - 1)]$
*Weibull survival to age (x)* $\quad\quad\;$ $= e^{\wedge}[-u * x^{(n+1)}]$

*Where A, m, c, and n are fitted variables for the current age of target individual.*

*Notice that the survival probability must be constrained to a boundary of 0 to 1. As a result, the power term of the exponent (which starts with "-") must be less that 0. So future age x is positive and therefore we can know some boundary information forced on variable "u".*

## Formula Z

| | |
|---|---|
| *variance of Y* | = *variance of* $[(X + \frac{1}{2}) * e^{\frac{1}{2}}]$ |
| | = $(e^{\frac{1}{2}})^2 *$ *variance of* $(X + \frac{1}{2})$ |
| | = $e^1 *$ *variance of X* |
| | = $(e) * (1/12)$ |
| | = $[f'(x_0)]^2 *$ *(variance of X)* |

## Formula AA

| | |
|---|---|
| $SD_Y$ | = *(slope)* $* SD_X$ |
| $slope^2$ | = $1.65^2$ |
| | $\approx 2.7$ |
| | $\approx e$ |
| *variance of Y* | = $SD_Y^2$ |
| | = $slope^2 * (SD_X)^2$ |
| | = $e *$ *(variance of X)* |
| | $\approx e * (1/12)$ |

## Formula AB

$$s_p^2 = \frac{(n_1 - 1)s_1^2 + (n_2 - 1)s_2^2}{n_1 + n_2 - 2}$$

## Formula AC

| | |
|---|---|
| $j$ | = merchant number |
| $y_j$ | = location of merchant$_j$ |
| $u$ | = Silk Road route length |
| $S_{sum}$ | = sum of all data(j) $[S(y_j)]^2[lnS^{**}(y_j)-lnS^{**}(y_{j+1})]$ |
| $F_{sum}$ | = sum of all data(j) $[F(y_j)]^2[lnF^{**}(y_{j+1})-lnF^{**}(y_j)]$ |
| $A^2$ | = $-n*F^{**}(u) + n*S_{sum} + n*F_{sum}$ |

Where ** means our hypothetical probability distribution model, and not the empirical sample distribution.

| | |
|---|---|
| $S_{sum}$ | $= [S(0)]^2[lnS^{**}(0)-lnS^{**}(3000)] + [S(3000)]^2[lnS^{**}(3000)-lnS^{**}(4000)] + [S(4000)]^2[lnS^{**}(4000)-lnS^{**}(5000)]$ |
| | $\approx 1^2[0+.69] + [½]^2[-.69+1.1] + [1/3]^2[-1.1+1.8]$ |
| | $\approx 0.87$ |
| $F_{sum}$ | $= 0 + [F(3000)]^2[lnF^{**}(4000)-lnF^{**}(3000)] + [F(4000)]^2[lnF^{**}(5000)-lnF^{**}(4000)] + [F(5000)]^2[lnF^{**}(6000)-lnF^{**}(5000)]$ |
| | $\approx 0 + [½]^2[-.41+.69] + [2/3]^2[-.18+.41] + [5/6]^2[0+.18]$ |
| | $\approx 0.30$ |
| $A^2$ | $\approx -3*F^{**}(6000) + 3*0.87+ 3*0.30$ |
| | $\approx 0.51$ |

# References

[i] An Intermediate Course in Probability, A. Gut
[ii] Fifty Challenging Problems in Probability, F. Mosteller
[iii] http://statisticalideas.blogspot.com/2014/03/briefer-economic-cycles.html
[iv] A First Course in Probability, S. Ross
[v] http://www.malaysiaairlines.com/my/en/site/dark-site.html
[vi] http://statisticalideas.blogspot.com/2014/03/briefer-economic-cycles.html
[vii] Probability and Random Processes, G. Grimmett
[viii]
http://www.princeton.edu/~sircar/Public/ARTICLES/VIXoptions101812.pdf
[ix] Statistical Security for Social Security, Soneji, King, Demography
[x] http://www.cbo.gov/publication/43524
[xi] ASOP No. 46, Risk Evaluation in Enterprise Risk Management, Actuarial Standards Board
[xii] Mathematical Statistics and Data Analysis, J. Rice
[xiii] Options, Futures, and Other Derivatives, J. Hull
[xiv]
http://www.census.gov/compendia/statab/2012/tables/12s0059.pdf
[xv] http://www.census.gov/prod/2011pubs/p70-126.pdf
[xvi]
http://www.census.gov/compendia/statab/2012/tables/12s0064.pdf
[xvii] http://en.wikipedia.org/wiki/Durbin–Watson_statistic
[xviii] A First Course in Probability, S. Ross
[xix] Financial Enterprise Risk Management, P. Sweeting
[xx] http://quickfacts.census.gov/qfd/states/48/4835000.html
[xxi] http://statisticalideas.blogspot.com/2014/03/probability-simulations-flight-mh370.html
[xxii] http://arxiv.org/abs/1403.5833
http://papers.ssrn.com/sol3/cf_dev/AbsByAuth.cfm?per_id=2245916

http://ideas.repec.org/cgi-bin/htsearch?q=salil+mehta

[xxiii] http://statisticalideas.blogspot.com/2013/12/probability-and-escape-velocity.html
[xxiv] Partial Differential Equations for Scientists and Engineers, S. Farlow
[xxv] Probability Models, S. Ross
[xxvi] Mathematical Statistics with Applications, A. Kapadia, W. Chan, L. Moyé
[xxvii] Mathematical Statistics and Data Analysis, J. Rice
[xxviii] http://statisticalideas.blogspot.com/2013/05/solving-quadnomial-probability.html
[xxix] Mathematical Statistics and Data Analysis, J. Rice
[xxx] Options, Futures, and Other Derivatives, J. Hull
[xxxi] http://statisticalideas.blogspot.com/2013/08/credibility-of-fibonacci.html
[xxxii] http://statisticalideas.blogspot.com/2013/09/attack-on-eastern-damascus.html
[xxxiii] Probability, Statistics, and Truth, R. von Mises
[xxxiv] http://statisticalideas.blogspot.com/2013/08/masked-statistical-recalibrations-in.html
[xxxv] http://www.federalreserve.gov/monetarypolicy/files/fomcprojtabl20120125.pdf
[xxxvi] http://statisticalideas.blogspot.com/2013/06/a-20-chance-of-rate-firming-prior-to.html
[xxxvii] Probability Models, S. Ross
[xxxviii] http://statisticalideas.blogspot.com/2014/01/n-law.html
[xxxix] [xxxix] http://statisticalideas.blogspot.com/2014/03/p-statistics-in-research.html
[xl] http://statisticalideas.blogspot.com/2014/03/probability-simulations-flight-mh370.html
[xli] http://www.cdc.gov/nchs/data/nhsr/nhsr010.pdf
[xlii] http://www.jlin7.com/
[xliii] http://www.bea.gov/newsreleases/national/gdp/2013/pdf/gdp2q13_adv.pdf

xliv

http://www.mckinsey.com/insights/high_tech_telecoms_inter
net/measuring_the_full_impact_of_digital_capital

xlv http://e.businessinsider.com/public/1852976

xlvi Advanced Engineering Mathematics, Kreyszig

xlvii Investment Science, D. Luenberger

xlviii Actuarial Mathematics, N. Bowers, H. Gerber, J. Hickman, D. Jones, C. Nesbitt

xlix http://statisticalideas.blogspot.com/2013/06/battery-warrant-valuations.html

l Calculus Early Transcendentals, J. Stewart

li Introduction to Mathematical Statistics, P. Hoel

lii Asset Valuation and Equity, CFA Institute

liii http://statisticalideas.blogspot.com/p/test.html

liv https://sites.google.com/site/statisticalideas/home/ch2

lv Actuarial Mathematics, N. Bowers, H. Gerber, J. Hickman, D. Jones, C. Nesbitt

lvi Loss Models: From Data to Decisions, S. Klugman, H. Panjer, G. Willmot

lvii http://statisticalideas.blogspot.com/2014/03/briefer-economic-cycles.html

lviii Discourse on Method, Optics, Geometry, and Meteorology, P. Olscamp

lix Value at Risk, P. Jorion

lx http://statisticalideas.blogspot.com/2013/09/tail-risk-have-we-met.html

www.ingramcontent.com/pod-product-compliance
Lightning Source LLC
Chambersburg PA
CBHW051806170526
45167CB00005B/1907